GRAND BATTERY

GRAND BATTERY

A Guide and Rules to Napoleonic Wargames

Jonathan Sutherland &
Diane Canwell

Pen & Sword
MILITARY

First published in Great Britain in 2011 by
Pen & Sword Military
an imprint of
Pen & Sword Books Ltd
47 Church Street
Barnsley
South Yorkshire
S70 2AS

ISBN 978 1 84415 941 3

A CIP catalogue record for this book is available from the British Library

Printed and bound in India by
Replika Press Pvt. Ltd.

Pen & Sword Books Ltd incorporates the Imprints of Pen & Sword Aviation, Pen & Sword Maritime, Pen & Sword Military, Wharncliffe Local History, Pen and Sword Select, Pen and Sword Military Classics, Leo Cooper, Remember When, Seaforth Publishing and Frontline Publishing.

For a complete list of Pen & Sword titles please contact
PEN & SWORD BOOKS LIMITED
47 Church Street, Barnsley, South Yorkshire, S70 2AS, England
E-mail: enquiries@pen-and-sword.co.uk
Website: www.pen-and-sword.co.uk

CONTENTS

Acknowledgements

The author would like to thank the following for providing images: Joe Dever, Pete Moreby (Elite Miniatures), Wargames Holiday Centre, Scarborough, Warmodelling, Garrison Miniatures, Front Rank Miniatures, Baccus, Miniature Figurines (Caliver Books), Eureka Miniatures, The Assault Group and Vinnie Singleton.

The editor would like to thank the members of Rainham Wargames Club for their time, expertise and figures used in playtesting, with special thanks to Graham Down and to Tim of Frontline Wargames for the photo shoots. Thanks also to Nick Dingle for additional playtesting and very useful comments.

INTRODUCTION

Welcome to the first of Pen & Sword's series of wargaming guides and rules. The series is much more than a set of wargame rules or a description of the weapons and tactics of the period. In this title you will find a brief history of the period, a primer on the weapons and tactics used by the protagonists and information about the uniforms and transport used.

The series is aimed at the wargamer, the newcomer to the hobby and the old hand with years of experience of building up miniature forces and playing wargames. The rules are designed to be easy to pick up and play, novel in their approach and above all, as close a representation of the real fighting as possible. It is never likely that we can reproduce what happened on a real battlefield, but the rules are designed to give a true flavour of the type of warfare without overcomplicating things and making the game play ponderous to the point that they are unplayable.

Wargaming is a social activity in the main and the rules assume that the hours that the wargamer has spent alone organising, painting, basing and protecting the figures and vehicles will be rewarded by a game with fellow players. That said, it is perfectly possible to play against a programmed 'dummy' opponent who will try to react to your actions and will attempt to mimic the tactics and strategy used by a real enemy on the battlefield.

Grand Battery covers the broad sweep of the Napoleonic Wars, essentially a series of wars fought between France and other nations between 1799 and 1815. We arbitrarily choose this period as 1799 was the year in which France came under the domination of Napoleon Bonaparte. The guide covers a wide variety of different nations, from the Austrians to the Ottoman Turks and from the British to the innumerable German states. The scenarios in this book are designed to provide a taste of gaming the period with three battles from 1793-1815,

The guide also includes a unique buyer's guide which details most of the manufacturers of figures and vehicles for the period; additional sections cover terrain and buildings. Each set of rules in the series will be supported by a series of scenario books. Each of these focuses on a theatre of the war or a particular aspect of the warfare, such as the Peninsular War or Napoleon's campaigns in Egypt. Each scenario book will include at least twenty scaled scenarios with additional rules, new organisation charts and original photographs of the games in progress.

Pen and Sword also have online resources to support the series, at www.pen-and-sword.co.uk and there you will find additional scenarios, rules clarifications, battle reports, downloadable cards, templates and ready reference sheets and a host of other game support.

Have fun and good gaming!

Jon Sutherland and Diane Canwell

CHAPTER 1:

Period History, Chronology and Major Battles

Chronology of Napoleonic Battles

The Napoleonic Wars, although they only spanned around sixteen years, involved nearly every European nation. The wars also spilled out into North and South America and into Egypt. The Napoleonic Wars were primarily fought between France and her allies and a shifting coalition of nations. A constant against the French were the British and at various times they were in alliances with Russia, Austria, Portugal and other countries.

The Napoleonic Wars represent perhaps the first total war; in the sense that the wars were being fought by primarily conscripted soldiers and whole nations found themselves embroiled in the conflict rather than their professional armies.

The French managed to fashion a huge number of victories, some of which would knock nations out of the conflict for some months or years before they could recover. There were periods

of peace, but they were short-lived. Systematically the French under Napoleon attempted to dominate Europe and reshape the states as they saw fit.

The wars saw innumerable small skirmishes and engagements, but also vast battles with tens, if not hundreds, of thousands of men involved. In Napoleon's invasion of Russia in 1812 he took with him an army of half a million men. This chapter can only scratch the surface of the Napoleonic period. It begins with a chronology identifying the key events and battles. We then turn our attention to some of the key engagements of the Napoleonic Wars, leading up to the last throw of the dice by Napoleon and the French, which ultimately ended in the destruction of his army at Waterloo and his exile, which marked the end of the Napoleonic period.

Year	Date	Conflict
1792	20 April	France declares war on Austria, Prussia and Piedmont, beginning the War of the First Coalition
	23 August	Capitulation of the fortification of Longwy
	14 May	Russian troops cross the border and Poland fights to defend the constitution
	2 September	Capitulation of the fortification of Verdun
	20 September	Battle of Valmy
	23 September	The Austrians encircle Lille
	1 October	The Convention splits French forces in eight armies: North, Ardennes, Moselle, Rhine, Vosges, Alps, Pyrénées, and Interior
	3 October	Revolutionary troops occupy Basel
	4 October	Revolutionary troops occupy Worms
	27 October	Revolutionary troops enter Belgium
	6 November	Battle of Jemappes
	9 November	Revolutionary troops occupy the Palatinate

1793	1 February	France declares war on England and Holland 7 February France declares war on Spain
	18 March	Battle of Neerwinden
	22 March	Great Britain, Holland, Spain, Naples and the Holy Roman Empire declare war on France, in reaction to the murder of the French king
	28 June	Battle of Valenciennes
	August-September	Siege of Toulon
	15/16 October	Battle of Wattignies
	28/30 November	Battle of Kaiserslautern
	22 December	Battle of Fröschwiller
	23 December	Battle of Savernay
	26 December	Battle of Geisberg
1794	24 April	Battle of Villers-en Cauches
	18 May	Battle of Tourcoing
	28 May - 1 June	Battle of Ushant
	26 June	Battle of Fleurus
	4 November	Kléber recieves the capitulation of Maastricht
	6 November	The Russians enter Warsaw
	8 November	French troops enter Nimwegen
	November/December	Civil war breaks out in the Vendée
	14 December	Kleber begins the siege of Mayence
1795	20 January	French troops occupy Amsterdam
	5 April	Peace treaty of Basel
	23 June	Battle of Lorient
	16 – 20 July	Battle of Quiberon

	22 July – 16 August	Battle of Cape of Good Hope
	29 August	Peace treaty between France and Sweden
	28 September	Treaty between Great Britain, Russia and Austria
	1 October	Annexation of Belgium by France
	24 November	Battle of Loano
1796	2 March	Bonaparte is named supreme commander of the Italian Army
	10 April	Beginning of Bonaparte's campaign in Italy
	12 April	Battle of Montenotte
	13 April	Battle of Millesimo
	14 – 15 April	Battle of Dego
	21 April	Battle of Mondovi
	10 May	Battle of Lodi
	15 May	Bonaparte enters Milan
	20 May	Military campaign in Germany resumed
	4 June	First siege of Montua
	16 June	Battle of Wetzlar
	9 July	Battle of Malsch
	16 July	Kleber enters Frankfurt
	18 July	Moreau enters Stuttgart
	25 July	Jourdan takes Würzburg
	3 August	Battle of Lönato
	5 August	Battle of Castiglione
	7 August	Kleber victorious against the Austrians at Forchheim
	18 August	Treaty between France and Spain: offensive and defensive alliance

	24 August	Battle of Amberg
		Battle of Friedberg
	3 September	Battle of Würzburg
	5 September	Battle of Caliano
	8 September	Battle of Bassano
	12 November	Battle of Caldiero
	15 – 17 November	Battle of Arcola
1797	14 – 15 January	Battle of Rivoli
	14 February	Battle of Cape St Vincent
	23 March	Battle of Malborghetto
	18 April	Battle of Lahn
	Summer	Napoleon's conquest of Venice
	11 October	Battle of Camperdown
	30 December	French troops enter Mayence
1798		Irish Rebellion
	19 May	Departure of Bonaparte's expedition to Egypt from Toulon
	June	Occupation of Malta, landing in Alexandria (July 2), conquest of the Nil region, push to Syria
	21 June	Battle of Vinegar Hill

	21 July	Battle of the Pyramids
	23 July	Bonaparte conquers Cairo
	1 August	Battle of Aboukir Bay
		Turkey declares war on France
	22 August	Battle of Killala Bay
	25 August	Battle of Castlebar
1799	10 February	Bonaparte leaves Cairo and leads his army towards Syria
	19 March – May	Siege of St Jean d'Acre
	25 March	Battle of Stockach
	5 April	Battle of Magnano
	17 April	Battle of Mount Tabor
	27 April	Battle of Cassano
	4 – 7 June	First Battle of Zurich
	17 – 19 July	Battle of Trebbia
	25 July	First Battle of Aboukir
	15 August	Battle of Novi
	16 September – 2 October	Battle of Bergen-op-Zoom
	25 September	Second Battle of Zurich

	6 October	Battle of Castricum
1800	20 March	Battle of Heliopoli
	20 April – 4 June	Siege of Genoa
	3 May	Battle of Stockach
	5 May	Battle of Möskirch
	16 May	First Battle of Ulm
	9 June	Battle of Montebello
	14 June	Battle of Marengo
	19 June	Battle of Hochstadt
	3 December	Battle of Hohenlinden
	24 December	Assassination attempt on Bonaparte
1801	9 February	Treaty of Lunéville
	21 March	Second Battle of Aboukir (Alexandria)
	2 April	Battle of Copenhagen
	6 – 12 July	Battle of Algericas
	8 October	Peace treaty between France and Russia
1803	26 May	French troops enter Hanover
	14 June	French forces enter Naples to occupy Puglia.
	15 June	French army masses at Boulogne for the invasion of England.
	21 June	British forces capture Santa Lucia

	30 June	British forces capture Tobago
	11 September	British General Lake defeats Mahrathas
	20 September	British capture Demerara and Essequibo
	23 September	Wellington defeats Mahrathas at Assaye.
	1 November	Lake defeats Mahrathas at Laswari
	28 November	Wellington defeats Mahrathas at Argaon
1805	8 – 21 October	Second Battle of Ulm
	21 October	Battle of Trafalgar
	26 October	Grande Armée marches from Boulogne to the Rhine
	30 October	Battle of Caldiero
	12 November	French occupy Vienna
	16 November	Battle of Oberhollabrunn
	2 December	Battle of Austerlitz
	12 December	Spain declares war on Great Britain
1806	6 January	British expeditionary force disembarks at Capetown
	8 January	Dutch defeated at Blauewberg
	18 January	Dutch surrender at Cape Province
	19 January	Anglo-Russian expeditionary force embarks from Naples

6 February	Duckworth destroys French fleet off Santo Domingo
9 February	French invade Naples
13 February	British evacuate Hanover
4 March	French besiege Gaeta
5 March	Russians seize Cattaro
9 March	Neapolitans defeated at Campo Tenese
26 May	French occupy Ragusa
17 June	Russians and Montenegrins defeat French outside Ragusa
18 June	Russians and Montenegrins besiege Ragusa
25 June	First Battle of Buenos Aires
4 July	Battle of Maida
12 August	Criollo militia recapture Buenos Aires
1 October	Russians repulse French assault at Castelnuovo
10 October	Battle of Saalfeld
14 October	Battle of Jena and Auerstadt
17 October	Prussians defeated at Halle

	22 October	French besiege Magdeburg
	25 October	Grande Armée enters Berlin
	28 October	Hohenlohe capitulates at Prenzlau
	29 October	French capture Stettin
	7 November	Battle of Lübeck
	28 November	French enter Warsaw
	24 December	Russians occupy Bucharest
	26 December	Battle of Pultusk
	29 December	Serbs capture Belgrade
1807	30 January	French besiege Stralsund
	3 February	British capture Montevideo; Bennigsen repulses Napoleon at Jenkendorf
	8 February	Battle of Eylau
	18 March – 27 May	Siege of Danzig
	21 March	British capture Alexandria
	29 March	British defeated in First battle of Rosetta
	21 April	British defeated in Second Battle of Rosetta
	2 June	Turks recapture Bucharest
	10 June	Battle of Heilsberg
	13 - 14 June	Battle of Friedland
	1 July	Russians defeat Turkish navy off Lemnos
	5 July	Second Battle of Buenos Aires
	16 July	British expeditionary force lands on the Island of Rügen
	3 August	French occupy Ionian islands
	16 August – 20 October	Siege of Copenhagen

	17 August	Turks besiege Alexandria
	14 September	British evacuate Alexandria
	19 October	French forces enter Spain en route for Portugal
	30 November	French forces enter Lisbon
1808	2 February	French occupy Rome
	9 February	French troops enter Catalonia
	16 February	French seize Pamplona
	21 February	Russia invades Finland
	5 March	French troops seize San Sebastian
	23 March	French occupy Madrid
	16 May	France annexes Rome
	17 May	British expeditionary force arrives at Göteborg and garrison capitulates
	15 June	French open first siege of Zaragoza
	19 July	Battle of Baylen
	17 August	Battle of Rolica
	21 August	Battle of Vimeiro
	25 August	British defeat Russians in naval action off Hango
	29 October	Lefebvre defeats Blake at Zornoza
	5 November	Bonaparte enters Spain
	7 November	Spaniards defeat French garrison of Santo Domingo at Palo Hincado
	10 November	Bonaparte defeats Belveder at Gamonal
	10-11 November	Victor defeats Blake at Espinosa de los Monteros
	13 November	British advanced guard reaches Salamanca
	23 November	Lannes defeats Castanos at Tudela
	30 November	Bonaparte defeats Spaniards at Somosierra
	4 December	Madrid surrenders to Bonaparte
	16 December	Saint Cyr relieves Barcelona

	20 December	French besiege Zaragoza for a second time
	21 December	Battle of Sahagun
	23 December	Moore retreats to Corunna
	26 December	Battle of Benevente
1809	7 January	British and Portuguese forces seize Cayenne
	13 January	Victor defeats Venegas at Ucles
	16 January	Battle of Corunna
	30 January	British invade Martinique
	19 February	Zaragoza surrenders
	24 February	Martinique capitulates
	25 February	Saint Cyr defeats Reding at Valls
	9 March	Soult invades Portugal
	20 March	Portuguese defeated at Braga
	28 March	Battle of Medellin
	11 – 12 April	Battle of Basque Roads
	15 April	Austrians invade Grand Duchy of Warsaw under Archduke Ferdinand
	16 April	Battle of Sacile
	19 April	Ferdinand defeats Poles at Raszyn
	20 April	Battle of Abensberg
	21 April	Battle of Landshut
		Austrians occupy Warsaw
	22 April	Battle of Eckmühl
	12 May	Battle of Douro (Oporto)
	21 – 22 May	Battle of Aspern-Essling
	22 May	French defeated by Spaniards at Santiago
	23 May	Blake defeats Suchet at Alcaniz
	8 June	Spaniards defeat French at Sanpayo
	11 June	Brunswick seizes Dresden
	14 June	Battle of Raab
	15 June	Suchet defeats Blake at Maria
	18 June	Suchet defeats Blake at Belchite
	19 June	Brunswick seizes Leipzig
	3 July	Wellington enters Spain
	5 – 6 July	Battle of Wagram
	6 July	French garrison of Santo Domingo surrenders
	15-17 July	Poniatowski captures Cracow and defeats Austrians at Wieniawka
	27 - 28 July	Battle of Talavera
	2 – 16 August	British besiege Flushing
	30 September	British recapture Ionian islands
	18 October	Del Parque defeats French at Tamames

	19 November	Battle of Ocana
1810	27 January	British invade Guadeloupe
	31 January	Soult captures Seville
	5 February	Siege of Cadiz
	21 March	French besiege Astorga
	22 March	British invade Santa Maura (Lefkas)
	29 March	French occupy Oviedo
	13 April	French besiege Lerida
	22 April	French capture Astorga
	14 May	French capture Lerida
	15 May	French besiege Mequinenza
	6 June	French besiege Ciudad Rodrigo
	8 June	French capture Mequinenza
	1-2 July	Siege of Ciudad Rodrigo
	9 July	First Battle of Ciudad Rodrigo
	21 July	Massena invades Portugal
	24 July	British rearguard driven over River Coa
	27 August	Massena captures Almeida
	17-18 September	French attempt to invade Sicily is driven off
	27 September	Battle of Busaco
	29 November – 3 December	British invade and capture Martinique
	16 December	Suchet besieges Tortosa
1811	5 March	Battle of Barrosa
	11 March	First Battle of Badajoz
	3 April	Battle of Sabugal

	7 April	Wellington blockades Almeida
	3 - 5 May	Battle of Fuentes de Oñoro
	6 May	Wellington besieges Badajoz
	8 May	Suchet besieges Tarragona
	12 May	Siege of Badajoz
	16 May	Battle of Albuera
	17 June	Wellington retreats into Portugal
	23 June	Spaniards defeat French at River Orbigo
	4 August	British invade Java
	11 August	Wellington blockades Ciudad Rodrigo
	19 August	French capture Figueras
	18 September	Garrison of Java capitulates
	25 September	Marmont drives back Wellington at El Bodon
	28 October	Battle of Arroyo dos Molinos
	20 December	French besiege Tarifa
1812	9 January	Battle of Valencia
	8 - 19 January	Second Battle of Ciudad Rodrigo
	2 February	French capture Peniscola
	16 March - 6 April	Second Battle of Badajoz
	19 May	Battle of Almaraz
	15 June	French evacuate Oviedo
	28 June	French occupy Vilna (Vilnius)
	8 July	French occupy Minsk
	22 July	Battle of Salamanca
	23 July	Battle of Mogilev
	24 July	Macdonald besieges Riga
	28 July	French occupy Vitebsk
	12 August	Wellington enters Madrid
	13 August	Spaniards capture Bilbao
	17 – 19 August	Battle of Smolensk
	19 August	Battle of Valutino
	24 August	Soult raises siege of Cadiz and commences evacuation of Andalucia
	7 September	Battle of Borodino
	15 September	Napoleon enters Moscow as Russians evacuate
	19 September – 22 October	Siege of Burgos
	18 October	Murat defeated at Vinkovo
	24 – 25 October	Battle of Maloyaroslavets
	31 October	British evacuate Madrid
	3 November	Grande Armée attacked at Fyodorovskoy
	16 – 17 November	Battle of Krasnyi

	26 – 28 November	Battle of Berezina
1813	4 January	French evacuate Königsberg (Kaliningrad) and retreat to the line of the Vistula (Wisla)
	8 January	Yorck occupies Königsberg
	12 – 16 January	Russian besiege Danzig
	4 March	Russians enter Berlin
	12 March	French evacuate Hamburg
	18 March	Prusso-Russian raiding force seizes Hamburg
	27 March	Prussians occupy Dresden
	2 April	French defeated at Lüneberg
	3-5 April	French defeated at Mockern
	13 April	French defeated at Second Battle of Castalla.
	18 April	Russians capture Thorn (Torun)
	21 April	Russians capture Spandau
	2 May	Battle of Lützen
	8 May	Napoleon occupies Dresden
	20 – 21 May	Battle of Bautzen
	31 May	Franco-Danish forces reoccupy Hamburg
	2 June	British launch amphibious operation against Tarragona
	18 June	British abandon operations against Tarragona
	21 June	Battle of Vittoria

	28 June	Wellington besieges San Sebastian
	25 July	Soult launches counter-offensive in the Pyrenees.
	26 July	Wellington raises siege of San Sebastian
	26 July – 1 August	Battle of Sorauren
	22 August	Wellington resumes siege of San Sebastian
	23 August	Battle of Grossbeeren
	26 August	Battle of Katzbach
	26 – 27 August	Battle of Dresden
	30 August	Battle of Kulm-Priesten
	31 August	San Sebastian falls to Wellington
	6 September	Battle of Dennewitz
	15 September	Allies besiege Magdeburg
	3 October	Battle of Wartenberg
	16 – 19 October	Battle of Leipzig
	30 October	Battle of Hanau
	5 November	Austrians besiege Venice
	9 November	Eugene defeats Austrians at Ala
	10 November	Battle of Nivelle
	9 – 13 December	Battle of Nive
	29 December	Allies move across the Rhine in force
1814	28 January	Murat invades Italy
	29 January	Battle of Brienne
	1 February	Battle of La Rothière
	8 February	Eugene defeats the Austrians at the River Minico
	10 February	Battle of Champaubert
	11 February	Battle of Montmirail
	12 February	Battle of Château-Thierry

	14 February	Battle of Vauchamps
	18 February	Battle of Montereau
	27 February	Battle of Orthes
		Battle of Bar-sur-Aube
	7 March	Battle of Craonne
	8 March	Battle of Bergen-Op-Zoom
	9 – 10 March	Battle of Lâon
	13 March	Battle of Rheims
	20 – 21 March	Battle of Arcis-sur-Aube
	25 March	Battle of La Fère-Champenoise
	30 March	Battle of Montmartre
	10 April	Battle of Toulouse
	12 April	Murat defeats French at Borgo San Donnino
	14 April	British repulse sortie by garrison of Bayonne
1815	1 March	Napoleon lands in France
	20 March	Napoleon enters Paris
	2 May	Battle of Tolentino
	15 June	Napoleon invades Belgium

16 June	Battle of Quatre Bras
	Battle of Ligny
18 June	Battle of Waterloo
	Battle of Wavre
19-21 June	Wellington and Blücher cross the frontier in pursuit of Napoleon's army
29 June	French defeated at La Souffel
29 June – 3 July	Allies surround Paris
1 July	Blücher occupies Versailles
10 July	Napoleon surrenders to British at Rochefort

In this part of the chapter we look at some of the key engagements. Some can be easily converted into wargames scenarios, whilst others are vast, sprawling battles that are best played as discrete

sections, focusing on particular key episodes within the battles. The battles listed are in date order. The Peninsular War battles are detailed as distinct wargame scenarios in both Chapter 8 of this book (where three have been selected) and in the accompanying *Wargames Scenarios: The Peninsular War 1808-1814*.

Montenotte, Battle of (1796)

Battle Name	**Dates**	**Campaign**
Montenotte	12 April 1796	1st Italian Campaign
Troops Involved		
France	9,000 880 losses	Commanded by Napoleon
Austria	6,000 2,500 losses (including prisoners)	Commanded by Argenteau

Outcome:
French victory

The Austrian and Piedmontese Allies were mutually distrustful of one another; Beaulieu commanded the Austrian army and Colli the Piedmontese. They maintained separate lines of communication back to Mantua (the Austrian administration centre) and to Turin (the Piedmontese capital). Whilst the Allies were thus divided, the French could take advantage of the situation and launch an offensive against one of them.

The blow fell at Montenotte on 12 April 1796. La Harpe's 9,000 troops advanced through thick fog in the early hours of the morning to close with Argenteau's 6,000 troops positioned at Montenotte. The most significant aspect of this whole operation, of which Montenotte was the first engagement, is that this was Napoleon's first independent command. He had assumed control of the Army of Italy on 27 March, inheriting a disjointed and ragged army which he would mould into an efficient and proud force.

La Harpe's troops split into two columns, with La Harpe heading straight for Mount Pra and Rampon, some way to the southeast. They would converge on the main Austrian position whilst a brigade of Meynier's cavalry, under Massena, would work around the Austrian right flank having passed through the Col di Cadibona.

Argenteau knew nothing of this until the fog lifted and he found himself confronted and outflanked. He immediately ordered a withdrawal to Montenotte Superior, but was beaten back by Massena's men. The garrison holding the Bric Castlas, at the centre of the Austrian position, was overwhelmed and Argenteau lost control of his troops as they routed.

This was Napoleon's first victory, his first demonstration of manoeuvre to bring a force into position to gain a local superiority and crack the enemy's resolve by striking hard at that point. It was a technique that would take some years to perfect, but he had begun.

Napoleon followed up the victory at Montenotte by attacking Colli. On 13 April he seized Dego then drove Colli all the way to Mondovi, where he beat him on 21 April. Just two days later, Colli sued for peace and on 28 April Piedmont was out of the war.

Beaulieu was similarly handled and at Lodi Napoleon managed to establish his reputation with his men. Five days after Lodi, Napoleon entered Milan in triumph. The 'little corporal' had been born.

Lodi, Battle of (1796)

Battle Name	**Dates**	**Campaign**
Lodi	10 May 1796	1st Italian Campaign
Troops Involved		
France	40,000 2,000 losses	Commanded by Napoleon
Austria	24,000 2,000 losses	Commanded by Beaulieu

Outcome:
French victory

The Austrian army had retired beyond the River Po and this was to afford Napoleon a glorious opportunity to prove his theories of manoeuvre. Ordering his army to concentrate around the towns of Tortona, Alessandria and Valenza, he intended to cross the Po and bring the Austrians to battle on his ground and at the time of his choosing.

He had three choices of crossing; Valenza would place the lead elements of his army at a considerable disadvantage as it would face the main Austrian force. Success was unlikely here as it was also at the crossing to the south of Pavia, again within Austrian striking distance. Remarkably, he chose Piacenza, 50 miles west and well in the Austrian rear. This would place his army between Beaulieu and his supplies and the only route of escape.

Napoleon mounted a diversionary attack on the Valenza crossing. Dallemagne, with 3,600 Grenadiers and 2,500 cavalry, would seize the Piacenza Bridge. The south banks of the River Po would be covered by other French troops, thus masking the manoeuvre. At 0400 on 7 May, Dallemagne left Stradella and reached Piacenza by 2100. It was fortunate that he had been reinforced by La Harpe's division, as on 4 May Beaulieu had sent Liptay, with a division of infantry supported by cavalry, to cover the crossings at Pavia. Beaulieu himself, after the attacks at Valenza, had decided to pull the bulk of the army across the River Ticino. Beaulieu now knew of the attempt on Piacenza and was sending Wukassovitch, with 4,500 men, from Valeggio to cover it.

On 8 May, Dallemagne began to force the Piacenza crossing, facing Liptay's troops who were thrown back to Fombio. The next day Dallemagne and La Harpe attacked and pursued Liptay's troops for two hours. Meanwhile, Augereau's troops had crossed the Po further west, at Varetto, and Massena and Serurier were close.

Beaulieu was retreating on the River Adda, but the leading French units were in sight of Lodi on the river by the early hours of 10 May. By this time nearly all of the Austrians had crossed and the Austrian rearguard consisted of 10,000 men under Sebottendorf.

The French cleared the town quickly and, after two assaults on the bridge, the French were over the river and the Austrian rearguard were running. They had done their job; Beaulieu and the Austrians had evaded an audacious plan. Napoleon would, however, perfect the technique in the years to come.

Castiglione, Battle of (1796)

Battle Name	**Dates**	**Campaign**
Castiglione	5 August 1796	1st Italian Campaign
Troops Involved		
France	27,500 1,500 losses	Commanded by Napoleon
Austria	25,000 3,000 killed or taken prisoner	Commanded by Würmser

Outcome:
French victory

Defeating the 12,700 Austrian garrison of Mantua would unlock their control of northern Italy and the siege, beginning on 3 June 1796, would dominate proceedings for the next eight months.

By July, however, two Austrian armies were marching to relieve the garrisons, forcing Napoleon to order the abandonment of the siege in a desperate attempt to concentrate his forces. Fortunately for the French, the Austrian advance was slow, but it was beginning to gather momentum as they forced Massena to abandon Verona. The two Austrian armies were not yet united and in this Napoleon still had a chance to deal with them.

On 3 August, Massena moved against the smaller of the Austrian armies at Lonato. Quasdanovich, the Austrian commander, suffered heavily and was thrown back in disorder. Napoleon now determined to deal with Würmser, who, if allowed to reach Mantua, would be sufficiently strong to beat his forces in northern Italy.

Napoleon began to concentrate his forces on the plain of Castiglione. Massena and Augereau, supported by Kilmaine's cavalry, would occupy Würmser frontally with an assault. Hopefully, Despinois, moving up from Brescia, would assist in this attack. Napoleon's ace was Sérurier's division, which he hoped would hit the left rear of the Austrians and cut off the enemy's retreat.

Unfortunately, not everything went according to plan. Sérurier was ill and the Corsican commander, Fiorella, took over his division and launched the flank attack prematurely. Napoleon's initial rearward movement to encourage Würmser to advance was in progress, but now he had to launch the assault to tie down the Austrians.

Napoleon urged his men forward with Massena and Augereau's troops striking the main Austrian line between Solferino and Monte Medolano. Marmont, still a low-ranking officer, was given command of the artillery, delivering telling bombardment on the left of the Austrian line in prelude to the infantry assaults, led by the grenadiers.

Würmser's right collapsed as Despinois' men finally arrived and stormed through the Austrian line. Würmser, ordering a general retreat, managed to extricate the bulk of his army in the nick of time, saving it from the fate Napoleon had planned for it.

Napoleon's troops had entered the combat exhausted and after fighting the engagement pursued for some three hours before it was called off.

The Austrians' first attempts to relieve Mantua had failed and for the time being French control of northern Italy was secured. The Austrians, however, would surely try again.

Arcola, Battle of (1796)

Battle Name	**Dates**	**Campaign**
Arcola	15 – 17 November 1796	1st Italian Campaign
Troops Involved		
France	Day 1: 15,000 Day 2: 17,000 Day 3: 20,000 4,500 - 4,600 losses	Commanded by Napoleon
Austria	Day 1: 3,000 – 9,000 Day 2: 12,000 Day 3: 17,000 6,000 – 7,000 losses	Commanded by Alvintzi

Outcome:
French victory

With the Austrian garrison at Mantua bolstered by troops from Würmser's army, Napoleon renewed his siege on 24 August. In September, however, he left 8,000 men to cover the city and marched north to Trent. The Austrians were again mustering to relieve Mantua and Napoleon defeated Davidovich at Caliano on 4 September, driving him through Trent. Napoleon then defeated Würmser at Bassano on 8 September. The second Austrian attempt to relieve Mantua had failed.

By October, the Austrian garrison at Mantua had reached 28,000, held by just 9,000 French. A new attempt to relieve the city was launched by Alvintzi, who was moving on Vicenza, hoping to meet with Davidovich at Verona. Napoleon sent Massena and Angereau to hold up Alvintzi, but on 12 November they encountered his whole army at Caldiero and were pushed back. Napoleon then despatched Vaubois to hold off Davidovich whilst he moved to engage Alvintzi.

On 14 November, Napoleon advanced along the south bank of the River Adige and crossed the river at Ronco, some 4 miles south of Arcola. Alvintzi still advancing on Arcola, sent reinforcements to the Austrian garrison at Arcola and more troops to deal with Massena at Porcile (Massena was protecting Napoleon's left flank). In turn, Napoleon sent Guieu via Albaredo to attack Arcola from the east.

On 15 November, Massena held off Provera's men whilst Napoleon's main force attempted to force the bridge at Arcola. Even as Napoleon's troops were gaining the upper hand, news came that Vaubois was falling back. It was not clear how badly he had fared, but Napoleon called off the attacks, allowing the Austrians to reinforce Arcola and occupy Porcile.

On 16 November, Napoleon launched more attacks on the bridge but failed to take it. After a day's struggle, Porcile was finally taken. It was now clear that Alvintzi's main force was in retreat and renewed attacks on 17 November aimed to cut off what remained of the Austrians. Massena occupied the Austrians at Arcola whilst Angereau was ordered to swing east and capture the town from the rear. Massena managed to storm the Arcola Bridge, meanwhile troops were sent across the River Alpone to distract the Austrians, allowing Angereau to advance north towards Arcola. Alvintzi now ordered a full withdrawal to Vicenzia for fear that his whole army would be surrounded and beaten by the French. Another attempt to relieve Mantua had failed, forcing Davidovich to retire also, but in the New Year another attempt would be made.

Pyramids, Battle of (1798)

Battle Name	**Dates**	**Campaign**
Embabeh	21 July 1798	Egyptian Campaign
Troops Involved		
France	25,000 300 losses	Commanded by Napoleon
Mamelukes	21,000 3,000 – 5,000 losses	Commanded by Murad Bey and Ibrahim Bey

Outcome:

French victory

Egypt was controlled by the Mamelukes, vassals of the Ottoman Empire, the country and army being led by Ibrahim Bey and Murad Bey. Napoleon had landed in Egypt on 1 July 1798 at Alexandria and was soon en route to capture Cairo.

On 20 July, Napoleon received word that the Mameluke army was concentrating around 20 miles from Cairo. Napoleon marched his worn-out army, suffering from the heat and lack of water, to the village of Embabeh on the banks of the Nile. The Mamelukes had foolishly allowed half of their forces to be caught still on the opposite bank of the Nile and for the whole battle Ibrahim's contribution could only be to sit and watch. Murad Bey had deployed his questionable *fellahin* (poor quality and poorer-armed infantry) in a defensive position protecting Embabeh. The main strike arm was his 6,000 mounted troops. Napoleon advanced against the Mamelukes in five divisional squares, with Bon and Vial on the river bank, Desaix and Reynier in the desert and Dugua in the centre, to act as a reserve.

Desaix detached troops to capture the village of Biktil and at around 1530 the Mamelukes began to mass their cavalry to strike at Desaix and Reynier. The two divisional squares only just managed to form up as the Mamelukes charged at them. The three columns of Mameluke cavalry swarmed around the Frenchmen, unable to break into its squares, constantly under artillery and musket fire.

The Mameluke cavalry swirled around, now under fire from howitzers in the middle of Dugua's square (in which was also Napoleon himself). Then they galloped back to try to deal with Desaix's detached troops, who had occupied Biktil. The Frenchmen clambered onto the rooftops to avoid being ridden down and sniped at the enemy cavalry.

Napoleon now launched Vial and Bon towards Embabeh, covered by his gunboats on the Nile. With the Mameluke cavalry occupied around Biktil, the French infantry fell on the impressed Mameluke infantry and in a matter of minutes they were within the defence works. Resistance was futile and brief and the *fellahin* began to flee (some 1,000 were drowned trying to cross the Nile).

By around 1630, the battle was effectively over and the Mameluke cavalry were in full retreat. Although the Mamelukes were decisively beaten, they would continue to harass Napoleon's troops. The French entered Cairo on 22 July, but they would not have very long to savour their victory, particularly after their fleet was destroyed at Aboukir Bay by Nelson on 1 August.

Marengo, Battle of (1800)

Battle Name	**Dates**	**Campaign**
Marengo	14 June 1800	2nd Italian Campaign Marengo Campaign
Troops Involved		
France	23,000 7,000 losses	Commanded by Napoleon
Austria	28,500 14,000 losses	Commanded by Melas/Zach

Outcome:
French victory

The Austrian general Melas was centred round Alessandria and found his communications with Turin cut by the French. Ott's Austrians were attacked at Montebello on 9 June by Lannes and Victor and the whole of Napoleon's army could then advance towards Alessandria. They set up positions at Marengo. Napoleon sent Desaix to Novi to prevent Melas from escaping to Genoa and a further force north to hold the Po River crossings. Napoleon fully expected Melas to retire and was caught completely unawares when Melas launched a full-scale attack in the early hours of 14 June.

As the Austrians engaged Victor's division, Napoleon was still not convinced that Melas' attack was a full assault. By 1100 his mind was made up and he began to send increasingly-desperate pleas to his commanders to move up immediately before the Austrians overran his positions.

Victor's troops were giving ground, as Ott's Austrians struck towards Castel Ceriolo, Napoleon threw in his only reserve to stem the tide. At around 1200, the Austrians paused to reorganize, Melas confident that the battle had been won. By 1500, Melas handed over the command of the expected pursuit to Zach, returning to Alessandria to have a minor wound dressed. Fortune was not with the Austrians and to Napoleon's relief came Desaix. He had begun his march but had been impeded by floods. Hearing canons firing back at Marengo he literally marched to the sound of the guns.

His troops began arriving on the battlefield at around 1700 and Napoleon ordered him to counterattack immediately, despite the exhaustion of Desaix's men. Napoleon sent Marmont's artillery and Kellerman's cavalry brigade in support of Desaix's advance. Desaix was shot dead at the head of his troops and for a moment his men hesitated, but Marmont's artillery began blowing holes in Zach's Austrians.

The key moment came when an Austrian caisson exploded, the Austrian troops were stunned for a second, and at that moment Kellerman charged. Immediately, Zach's troops began to rout and panic now spread all along the Austrian line, giving the weary French troops new heart. At first, the Austrian retreat was orderly, but it soon degenerated into a desperate rout.

On the evening of 15 June, Melas reluctantly signed an armistice agreeing to evacuate Lombardy. On 3 December 1800, Moreau beat the Austrians once more at Hohenlinden, which forced the Austrians to sign the Treaty of Luneville (February 1801).

Alexandria, Battle of (1801)

Battle Name	**Dates**	**Campaign**
Alexandria	21 March 1801	Egyptian Campaign
Troops Involved		
France	20,000 4,160 losses	Commanded by Menou
Britain	14,000 1,468 losses	Commanded by Abercrombie

Outcome:
British victory

The British expeditionary force under Sir Ralph Abercrombie moved to deal with the substantial French garrison under General Menou at Alexandria. On the night of 20 March, Abercrombie's troops had established their lines across the isthmus with their right resting on the sea and the ruins of Nicopolis and their left on the lake of Aboukir and the Alexandria Canal.

The British troops faced the city from the southwest. Abercrombie posted Moore's division on the right; the Guards held his centre and three other brigades the left. In reserve were two infantry brigades and some cavalry.

No sooner had the camps begun to be established than the alarms were raised at 0300 as there was visible activity in the city. At 0330 the French came storming out and drove in Abercrombie's pickets. It seemed that the French, in their columns, intended to break the British line on Moore's portion of the position. The initial fighting fell on the Gloucestershire Regiment (28th) but the French attacks were repulsed. A second French column broke through and a vicious fight developed between the French and the Black Watch (42nd) in the ruins of Nicopolis.

By now, the 28th was surrounded (later the regiment would be awarded the distinction of wearing its badge on the front and back of its headgear), but the French were driven off when three fresh British units were thrown in. The 42nd, facing French cavalry, was under great pressure and it was at this point that Abercrombie received a mortal wound in the thick of the fighting. In the centre, the Guards drove off the French, whilst the British left were never under serious threat.

Continued attacks on the British right continued until around 0830, the 42nd alone had fought off two determined attacks by the French cavalry. As daylight illuminated the battlefield, British gunboats began to open up on the French, which in part, accounts for the disproportionate losses during the battle. The French columns seem to have suffered a great deal of disruption from these boats, added to a large naval gun which had been added to the battery close to the Gloucestershires.

Abercrombie died on 28 March of his wounds; three other guards and Moore had been wounded. After the battle, the city held out until August, at which point the whole of Egypt was given back to the Turkish Sultan Selim III.

Ulm, Battle of (1805)

Battle Name	**Dates**	**Campaign**
Ulm	8-21 October 1805	Ulm-Austerlitz Campaign
Troops Involved		
France	200,000 6,000 losses	Commanded by Napoleon
Austria	72,000 40,000 losses (of which 30,000 captured)	Commanded by Baron Mack von Liebereich

Outcome:
French victory

Napoleon was still in Paris on 13 September when he received news that Mack's Austrians had crossed the River Inn and were advancing on Munich. A week later, Murat confirmed that Mack had reached Ulm. Significantly, there was no sign of the Russian army heading to link up with the Austrians.

By 25 September, Napoleon had reached Nancy, the next day he was at Strasbourg. On the same day, 40,000 French under Lannes and Murat had left Strasbourg; their first objective was Freudenstadt and more importantly to entice the Austrians further west. Napoleon need not have concerned himself about the Austrians taking flight and heading east, they seemed rooted to the spot at Ulm.

By 3 October, the French army had reached a line between Stuttgart and Ansbach, with Ney holding the pivotal position at Stuttgart; the armies could now swing around the rear of Mack and head for the Danube. On 6 October, the French army was approaching Münster, Donaüworth, Neuburg and Ingolstadt on the River Danube. Still Mack remained stuck at Ulm and showed no intention of reacting. The crossing took place during 6 or 7 October. The French army now pivoted on Angsburg and movement was already underway to seize Landsberg and Memmingen, capturing the Austrian depots and preventing Mack from falling back towards Innsbruck.

The seizure of the crossings over the Rivers Lech and Isar effectively meant that the Russians were out of the equation; they would not reach Mack in time to provide any assistance.

The first real engagement occurred at Wertingen on 8 October, when nine battalions of Austrians were swept aside by Murat and Lannes. By the night of 11 October, Bernadotte was within 6 miles of Munich and Napoleon now engaged Murat to close the trap and strike at Mack with the minimum delay.

That day (11 October) did not, however, all go according to plan as far as the French were concerned. Murat was advancing towards Ulm on the south bank of the River Danube, sure that Mack's main army was directly ahead of him. Consequently, he ordered Ney to bring two of his three divisions over the Danube and join him. This left Dupont very exposed and when he found himself facing 25,000 Austrians, including 10,000 cavalry some distance to the east of Michelsberg Heights, he was in serious trouble.

In typical style, however, Dupont attacked the dithering Austrians and his 4,000 men held on to the village of Albeck (or Haslach) all day, before retiring to Brenz after dusk. Mack's army had missed the opportunity to break out, sever Napoleon's supply and communications and link up with the Russians.

Soult was now at Landsberg and Napoleon, with Murat, Lannes, Ney, Davoüt and the Imperial Guard at Augsburg. Napoleon ordered 'not one is to escape' to his commanders as the French began to march west.

By the morning of 13 October, Napoleon was near Ulm and as the day developed several engagements of varying size were fought. At Albeck, Murat joined Dupont to see off a determined attack by Werneck's corps. This was achieved only after Ney had stormed the Elchingen Bridge against 9,000 Austrians. They had partially demolished the structure, but, under fire, Ney personally organized the repairs and then stormed the town. Any chance Mack had of escaping now looked doomed.

Archduke Ferdinand had seen enough, despite orders from Mack he broke out of the ring with 6,000 men. Ferdinand's cavalry force was ruthlessly pursued by Murat and only a handful managed to reach Heidenheim, where Werneck had retreated after his engagement with Murat and Dupont.

The climax of the operations took place on 15 October. Ney stormed the Michelsberg field work, allowing the French artillery to be set up to bombard Ulm itself. Mack sued for a temporary armistice. He offered to surrender in eight days if the Russians had not arrived. Napoleon refused and offered him five days respite in the certain knowledge that Kutusov's Russians were still a hundred miles away.

On 19 October, Werneck was forced to surrender to Murat at Trochtelfingten. The French general then moved on to Neustadt, where he accepted the capitulation of 12,000 Austrians. In a day, he had taken 20,000 prisoners, including 7 generals, 200 officers and 120 guns.

With these two new disasters, Mack offered his surrender on 20 October. Some 27,000 Austrian troops piled their arms and Austria was effectively out of the war in just twenty-six days. Ulm was a stunning victory and bought at the lowest possible cost in terms of French casualties. Only 10,000 Austrians made good their escape from the disaster. As for Mack, he was later court-martialled and sentenced to twenty years imprisonment.

Napoleon could be more than pleased with the results of the 'Ulm Manoeuvre' as it has become known. What he could not have known on 21 October was that his fleet was being destroyed at Trafalgar by Nelson on the very same day that Mack surrendered.

Austerlitz, Battle of (1805)

Battle Name	**Dates**	**Campaign**
Battle of the 3 Emperors	2 December 1805	Ulm-Austerlitz Campaign
Troops Involved		
France	66,800 – 73,200 9,000 losses	Commanded by Napoleon
Austria and Russia	85,400 – 85,700 15,000 losses (plus 11 – 12,000 captured)	Commanded by Kutusov/Tsar Alexander 1/Francis 1

Outcome:
French victory

As early as 21 November, Napoleon had chosen the ground at which he would turn and face the Austro-Russian army overlooking the village of Pratzen with the towns of Krenowitz and Austerlitz in the distance. He decided that he would allow the Allies to occupy the Pratzen Heights, Napoleon's right would be weak and his left strong, with the bulk of his army hidden around Zurlan, a peak of 850 feet. At the right moment he would unleash his army against the Allies on Pratzen Heights and fall on the enemy's rear.

As far as the Allies were concerned, they believed that Napoleon would not offer battle until he had reached Vienna. The Frenchman was happy to let them remain deluded and simply observed the snail-like Allied advance, willing them to close. On 30 November, Napoleon gave the orders to evacuate the Pratzen Heights and lure the Allies westwards. They began to occupy the exact positions he had anticipated between 1500 on 30 November and 0200 on 1 December.

Meanwhile, the French had massed some 60,000 men and were ready to move into position. Davoût occupied the Raigern Abbey. As far as the Allies were concerned, the French left was thinly held, as was the line along the Goldbach Brook. The Allies proposed that Docturov would hit Tellnitz at 0700 on 2 December, Langeron would advance between Tellnitz and Sokolnitz and Prschibitschewski, and Kollowrat would take the Sokolnitz Castle. Once this had been achieved, Kienmaier would move to block the Vienna to Brünn road and Bagration would tackle the French left, linking with Kollowrat. Constantine's troops would be held back in reserve, to be thrown in where needed.

Matters would, however, be delayed, probably due to translation problems and the fact that the Allied administration was not very professional, a factor which would resurface and confuse later.

At 0630 on 2 December, the French left and centre units began to occupy their positions and just after Buxhöwden threw Kienmair at Tellnitz. After an hour's fighting, the village remained in French hands, despite the fact that Soult had sent cavalry in against Buxhöwden's right flank, who were driven off. At 0830, Docturov's Russians hit Legrand on the Goldbach Brook and forced them away. Buxhöwden now halted, waiting for Langeron to take Sokolnitz. Langeron had been held up by Lichtenstein, who had himself been blocked by Kollowrat.

Davoût's men had not left Raigern Abbey until around 0630. He then marched, as per orders, for Turas, but new orders directed him to march for Sokolnitz as soon as possible. Fraint's men moved to assist Legrand and managed to regain Tellnitz. Docturov and Kienmaier counterattacked and retook the village but were again ordered to halt by Buxhöwden.

By 0900, supported by Allied artillery, Langeron and Prschibitschewski had taken Sokolnitz and the castle, but by 1000 Davoût retook the area, after having marched north. The battle, however, was not being decided on this part of the front and at just after 0830 Napoleon had launched Soult at the unsuspecting Allies.

St Hilaire headed for the high ground above Pratzen and Vandamme for the peak to the north of the village. Kollowrat had been following Prschibitschewski on the left, with Miloradovich on the right, also sending two battalions into Pratzen itself. These two battalions were immediately overwhelmed. Kutusov was at hand and ordered Miloradovich to face his right flank and for the Austrians to form up on Miloradovich's left. For a moment St Hilaire's leading units were stopped, then his second brigade emerged, supported by artillery. Kollowrat charged, was driven back and then charged again, this time supported by Russians. Vandamme then settled the issue by sweeping through Miloradovich, whose men promptly fled, followed by the Austrians. Vandamme now swung around the peak.

Along the Olmütz Road, Lannes carefully engaged Bagration, waiting for Murat and Bernadotte to form up beside him and fill the line towards Vandamme. Bagration was no keener to engage, he awaited Lichtenstein to reach his positions alongside him. Constantine, however, saw the group and hurried forward, leading the Russian Guard. He had, in his haste, committed the entire Allied reserve, possessing now the village of Blasowitz and for now protecting Bagration's left flank.

Lichtenstein was now forced to work his way around Constantine to get to his ordered position. No sooner had he got in place than he was attacked by the French cavalry, his own cavalry beat them off and gave chase, straight into Lannes' infantry which had formed squares. With that, Kutusov told him to move south and he complied, leaving half of his men with Bagration.

Meanwhile, on the Allied left, Langeron and Prschibitschewski were still fighting for possession of Sokolnitz with Davoût. The area changed hands several times and remarkably, given the 400 per cent Allied manpower advantage in this region of the battlefield, no Allied progress was being made either at Sokolnitz or at Tellnitz.

In the centre, the Russian Guard found itself unsupported, partly as a result of Lichtenstein's movement and the attack being launched by Soult. Rivaud's French stormed and took Blasowitz; the Russian Guard counterattacked, but was brought to a halt and badly mauled here by Rivaud and Cafferelli. Kutusov ordered the Russian Guard to the plateau, but by now they came under severe pressure from Vandamme who was firmly established and aggressive. In a matter of minutes, the Russian Guard was streaming back towards Krzenowitz. A counterattack from the

Russian Guard cavalry shattered Vandamme's over-exposed troops and the French advance here was stopped.

Bernadotte had been ordered to move up on Vandamme's left, Bessiere at the head of the Imperial Guard cavalry saved Vandamme after several charges. In the ensuing chaos, Lichtenstein was also swept away as Rivaud and Drouet (Bernadotte's corps) advanced to the edge of the Pratzen Plateau, watching the Russians streaming through Austerlitz.

Meanwhile, in the north, matters had not gone well for Bagration. He had tried to turn Lannes' flank and had advanced as far as Bosenitz, but he was stopped dead by artillery. The Russian artillery, however, opened up and ripped through the French ranks until Lannes counterattacked. Despite numerous Allied cavalry charges, Caffarelli could not be stopped. Nansouty now charged on Murat's orders and soon after Bagration's cavalry, and those which Lichtenstein had left him, were fleeing towards Austerlitz.

Bagration began a fighting retreat towards Posoritz, holding off Murat, but a determined all-arms attack by Lannes broke the Russians, who routed as far as Raussnitz. Under cover of darkness, he would then head south and retreat via Austerlitz.

At 1300 St Hilaire attacked towards Sokolnitz Castle, while Vandamme assailed Augeld, with elements of Legrand's, Oudinot's and Beaumont's forces filling the gap between. The Imperial Guard, around this time, had reached Saint Anthony's Chapel. At the same time Davoût launched an all-out assault, the Allies were trapped, only able to flee between the ponds of Satschen and Menitz. Thousands ran the gauntlet and thousands fell. Kienmaier did his best to cover the flight; Docturov had tried to make a last stand at Tellnitz, but was swept away by sheer weight of numbers and plummeting morale.

Austerlitz had been an Allied disaster, fought on Napoleon's battlefield, never seriously threatened by a chaotic and inefficient Allied force which had not lacked the guts to fight.

Maida (St Eufemio), Battle of (1806)

Battle Name	**Dates**	**Campaign**
Maida (St Eufemio)	4 July 1806	N/A
Troops Involved		
France	Circa 8,000 4,000 killed or wounded	Commanded by Reynier
British	4,795 Circa 326 losses	Commanded by Stuart

Outcome:
British victory

On 1 July 1806, a composite force under Stuart began landing in the Bay of St Eufemio on the coast of Italy. The plan was to support the anti-French elements within the new French kingdom of Naples.

Barely had the British troops landed when the fleet opened fire on the French-held castle of St Almanthea, which was quickly reduced with 900 prisoners falling into British hands. The nine battalions with supporting artillery headed inland, through the forest, towards Maida, sweeping aside a blocking move by a French regiment and netting another 200 prisoners.

Stuart issued a proclamation inviting the Neapolitans to rise up against the French, which

allowed the creation of the Calabrian Free Corps. Meanwhile, Reynier, based at Reggio, was en route to intercept the invaders and the rebels. Reynier took up positions at Maida with 4,000 infantry supported by cavalry and guns and awaited the arrival of reinforcements under Monteleone, which would give him a total force of some 8,000 men.

Stuart's first objective was to prevent the junction of the two French forces. Consequently, on 4 July, he marched for Maida hoping to get there ahead of Monteleone. In this he failed and Monteleone took up positions on the French right.

The two armies closed, with the French quickly moving forward as Stuart's troops deployed into line. Kemp's light infantry spread out whilst the British artillery began to open fire. Kemp held the right wing and engaged the French at a hundred yards. To his left was Acland's brigade and to their left were a provisional battalion of grenadiers and the 27th Foot.

MacLeod's Highlanders stove in the French line held by the 42nd Regiment (Imperial Guard) and Kemp's light infantry charged as the French wavered and fled. Acland now reinforced with the 78th and the 81st and swept the French left wing aside.

Three hundred French Dragoons tried to stem the rout, but these men were slaughtered by steady British volleys. Reynier's right and centre withdrew in good order and he was pursued for 3 miles before the British stopped. Reynier retreated towards Crotona, still shadowed by the Calabrians and the Highlanders. Thousands of mountain men flocked to aid the struggle against the French and this almost forgotten campaign would continue for another 2 years.

By 1808, Reynier had regained control of most of southern Italy, other notable engagements

including the siege of Castle Scylla, which fell to Reynier in mid-January 1808 after a long and gallant defence under Robertson.

Jena - Austerstadt, Battle of (1806)

Battle Name	**Dates**	**Campaign**
Austerstadt	14 October 1806	Prussian Campaign Jena Campaign
Troops Involved		
France	27,000 – 28,800 7,000 losses	Commanded by Davoüt
Prussia	63,000 – 65,000 25,000 losses	Commanded by Brunswick
Outcome:		
French victory		
Jena	14 October 1806	
Troops Involved		
France	46,000 – 96,000 5,000 losses	Commanded by Napoleon
Prussia	38,000 – 53,000 25,000 losses	Commanded by Hohenlohe/Rüchel
Outcome:		
French victory		

When the Fourth Coalition (Britain, Prussia, Russia and Portugal) was formed against France in 1806, the Grand Armée was in southern Germany and Napoleon planned to invade Prussia. If Napoleon could strike quickly, the other Allies would not have sufficient time to come to Prussia's aid before it was too late.

Napoleon could field a force of around 180,000, against whom the Prussians could muster some 130,000. Napoleon launched his offensive from Bavaria on 8 October 1806, advancing on a front of some 30 miles in three columns, screened by cavalry. Soult, Ney and the Bavarians occupied the right, Lannes and Augereau the left and the centre consisted of the Imperial Guard, Bernadotte and Davoüt. Even travelling through the Thuringian Forest, the French were covering some 15 miles per day.

Napoleon had planned to bring the Prussians to battle near Leipzig, but they had not deployed in the area. First contact was made at Saalfeld (10 October) when Lannes defeated a small Prussian force. On 13 October, however, Lannes encountered a larger Prussian army at Jena.

Napoleon now ordered the whole army to head west and support Lannes, arriving himself at 1600. Napoleon believed this was the main Prussian army, but, in fact, he was mistaken. Brunswick and the main Prussian force were actually heading for Auerstadt (some 15 miles away) and what Lannes had encountered was a large covering force under Hohenlohe.

Still labouring under the assumption that this was the main Prussian army, Napoleon gradually began filtering troops into the area (they had been marching through the night and into the following morning). The first to come in support of Lannes was the Imperial Guard at 1000 on 14 October, by which time Augereau, Soult, Ney and the reserve cavalry were close. Napoleon had around 80,000 men at his immediate disposal.

Napoleon now despatched Davoüt and Bernadotte on what he believed to be an enveloping manoeuvre around the Prussians. At Jena, Napoleon organized the line with Lannes in the centre, Augereau on the left and Soult on the right and sent them forward against the Prussians. Hohenlohe had just 33,000 men at this time and was sending desperate messages to Rüchel for assistance; Rüchel's 15,000 were in the Weimow area.

In the meantime, as Lannes, Augereau and Soult advanced Ney decided to launch his own attack on Hohenlohe against the orders of Napoleon. His cavalry charged at the Prussians and was nearly surrounded by the enemy. Napoleon had to quickly deploy Lannes and the reserve French cavalry to save him. Napoleon's plans, for the time being, were ruined, but fortunately Hohenlohe failed to take advantage of the situation.

Hohenlohe's hopes seem to have been resting on the arrival of Rüchel and he consequently chose to fight a delaying action. His infantry stood defiantly, being shot to pieces for some 2 hours whilst the rest of the French army reorganized itself.

Napoleon finally got his troops in order before 1500 and immediately launched a full-scale attack on the Prussians. Hohenlohe had welcomed the arrival of Rüchel an hour before, but between them they could not withstand the assault. Soon after the advance got underway, the whole of the Prussian army was in full retreat, having suffered horrendous casualties and losing thousands of men as they surrendered to the French.

Whilst Napoleon concentrated on what he still believed to be the main Prussian army, Davoüt had, in fact, contacted the larger Prussian force under Brunswick. Davoüt and Bernadotte had been sent on the enveloping march, but Bernadotte had actually marched south and would not become involved in either of the battles of 14 October. Davoüt's corps of around 27,000 faced Brunswick's force of up to 65,000. Luckily for Davoüt, the Prussian attacks were very poorly coordinated and he was able to defeat each of the assaults in detail. So solid and determined was

his defence, that he was able to take the initiative and offensive at around 1100, when he broke the back of the Prussian resolve.

By this time, Brunswick and Mollendorf (his second in command) had both been mortally wounded. The Prussians found themselves commanded by their king, Frederick William III. The Prussian monarch promptly ordered a retreat, but not until after his troops had suffered at least 13,000 casualties and a similar number captured. Davoüt's corps had suffered, proportionately, heavy casualties, but was still intact and pressing the Prussian retreat when Bernadotte finally arrived on the battlefield. His sole contribution was to take over from the exhausted men of Davoüt's corps in the pursuit.

Coupled with the defeat at Jena, the Prussian army had been shattered; Prussia lay at the mercy of Napoleon. He marched into Berlin on 24 October and the last Prussian army, under Blücher, surrendered at Lübeck on 24 November. The 'invincible' Prussian army had been destroyed and it would be seven more years before they faced Napoleon on the battlefield.

Eylau, Battle of (1807)

Battle Name	**Dates**	**Campaign**
Eylau	8 February 1807	Prussian Campaign Jena and Friedland Campaign
Troops Involved		
France	71,000 25,000 losses	Commanded by Napoleon
Russia	67,000 15,000 losses	Commanded by Bennigsen

Outcome:
Draw

The Prussian army all but collapsed after the defeats at Jena-Auerstadt, King Frederick William had fled to Russia and Napoleon was heading for Warsaw. All that stood before him was Bennigsen and remnants of the Prussian army, supported by a Russian army under the aged Kamenskoi.

Bennigsen had some 35,000 at Pultusk on 26 December 1806, when Lannes and Davoût attacked them. The engagement was indecisive, Bennigsen losing some 5,000 for the loss of 8,000 French. Another engagement at Golymin between Murat and Angereau had been similarly indecisive and these two expensive battles convinced Napoleon that the campaigning season was over.

Poland was not the ideal resting place for an army in winter; supplies were, at best, unreliable – not that the French supply system had ever been anything but amateur.

As it was, Kamenskoi had the misfortune to be assassinated on 9 January and the full control of the Russo-Prussian army fell squarely onto Bennigsen. He was clearly determined to make the best use of this opportunity and by the end of January was making attacks on the outlying French positions. Ney was immediately accused of having provoked the Russians during one of his many forays to pick up supplies by raiding, but the enemy's movement did seem to offer Napoleon an opportunity, perhaps, to deal with them decisively before the spring.

Napoleon first had to send couriers to the various corps commanders, telling them to concentrate. Unfortunately, a copy of the orders and attack plans fell into the hands of Russian Cossacks, who immediately passed them on to Bennigsen. This information allowed him to escape from an indecisive battle fought at Ionkovo on 3 February.

With Napoleon in pursuit, Bennigsen turned to face him at Eylau on 6 February. He had some 67,000 men and 460 cannon; Napoleon had just 45,000 men and 200 guns. Close by, however, were Ney and Davoût, with a further 26,000 men. Bennigsen could only rely on 9,000 Prussian troops (under Lestocq), but these were engaged in operations against Ney.

On the afternoon of 7 February, Napoleon closed on Eylau, the battle beginning by accident when the French baggage train moved into Eylau, oblivious of the fact that it was held by the Russians. After the town had changed hands several times, Napoleon found himself in possession by nightfall.

Weather conditions overnight and early the following morning did not auger well for a major engagement. Napoleon was, in any case, facing the enemy without the troops of Davoût or Ney. It was Napoleon's intention to hold and counter any enemy attack until such time that he was reinforced by the other two corps. Soult occupied the left, Augereau the right, with Murat's cavalry held in reserve to the rear.

The predominantly-infantry line of Bennigsen was supported on each flank by cavalry, but he had placed two large batteries of guns in the centre to soften up the French or break up any attacks. As it was, Bennigsen's hand was forced at around 0900 when Soult launched a spoiling attack. This convinced him to throw his own troops at the French left, supported by the massed artillery. Although the Russians had a clear superiority, the French inflicted far higher casualties on the densely-packed Russian columns.

Whilst the struggle continued, the advanced units of Davoût's corps began to arrive; Bennigsen immediately moved to hold them in order to finish off Soult. In desperation, Napoleon threw Augereau's corps into a frontal attack on Bennigsen's centre. By now the driving snow caused heavy casualties from friendly fire compounded by the fact that they marched straight at a seventy-gun enemy battery.

Things looked utterly bleak at 1030; Augereau's corps had been shredded and the Russians were in full advance in the centre of the battlefield. Soult was still holding, but only just. Nearly all of the regiments in Augereau's corps had been decimated; one of the 14th failed to retire with the rest of the troops and was engulfed by the Russian steamroller.

Napoleon had but one option, which was to commit his cavalry under Murat. The 10,700 men had to turn the tide or the battle would be lost. With Murat at it's head, this was, arguably, one of the most glorious cavalry charges in military history. Murat's men swept through the Russian infantry, reformed in the rear and overwhelmed the Russian battery which had slaughtered Augereau's troops. Although Murat lost 1,500 men, Napoleon's army was saved.

A total French victory was snatched from Napoleon's grasp by the timely arrival of Lestocq and his Prussians. Finally, Ney arrived and began an assault on the left, but by now darkness was closing in. The encounter had been extremely bloody, but inconclusive. If Napoleon wished to claim a victory and success he could; he had, at least, stopped Bennigsen's Russian winter offensive.

Friedland, Battle of (1807)

Battle Name	**Dates**	**Campaign**
Friedland	13-14 June	Prussian Campaign/Jena and Friedland Campaign
Troops Involved		
France	80,000 8,000 – 10,000 lossess	Commanded by Napoleon
Russian	60,000 20,000 – 25,000 losses	Commanded by Bennigsen

Outcome:
French victory

Napoleon resumed his offensive against the Russians on 10 June 1807, but once again Bennigsen decided to seize the initiative and attack at Heilsburg. It was an indecisive and costly battle for both sides and Napoleon even decided to withdraw and threaten the Russian flank. Bennigsen retired toward Friedland, his forward supply base and the River Alle.

Napoleon followed, despatching troops to cut Bennigsen off from Königsberg, where he believed the Russians were retreating. Instead, Napoleon received word that Bennigsen was concentrating his forces around Friedland on 13 June. By the morning of 14 June Napoleon was severely outnumbered, but troops were arriving all morning. Without waiting for Murat and Davoût to move up, he decided to launch an attack that day, the anniversary of Marengo.

Bennigsen had not chosen his ground well. If pressed, the entire army would have to cross the River Alle at Friedland, no doubt with the French hard on their heels. The Russians occupied a vast 4-mile arc with their backs to the river; if disaster struck, the whole army could be lost.

Napoleon waited as long as he dared, lest the Russians should ship across the River Alle during the night. At 1730 a preliminary bombardment signalled the first assault. It was to be made by Ney on the Russian left. As the Russians were forced back to the river, the French came under fire from Russian batteries sited on the opposite bank; Ney's men were stopped for the time being.

Napoleon now committed Victor to attack in support of Ney, falling on the Russian cavalry which had just been sent in to support Bagration's left. By now, a mass of Russian troops were streaming back towards the single bridge. A remarkable artillery feat from Sénarmont (Victor's artillery commander) snarled the Russian cohesion. He leapfrogged his batteries forward, bombarding the milling house, freeing up Ney to continue his advance.

Bennigsen tried to relieve the pressure on his left by launching his right and centre at the French. The attacks faltered and he threw in the Imperial Guard, again to no avail. Luckily for the Russians, a ford was found to the north of Friedland which served most of his right wing, but by 1100 the battle was over and only a Russian battery on the opposite bank of the River Alle covered the disordered rout. Bennigsen had narrowly avoided complete disaster, but on 19 June the Russians sued for peace.

Rolica, Battle of (1808)

Battle Name	**Dates**	**Campaign**
Rolica	17 August 1808	Peninsular War
Troops Involved		
France	3,000 500 losses	Commanded by Delaborde
Britain	4,000 400 losses	Commanded by Wellington

Outcome:
British victory

This was the first major engagement between the British and the French in the Peninsular War. Wellington had landed with a hastily-organized army at Mondego Bay around 100 miles north of Lisbon. His force, including 1,700 Portuguese, amounted to some 15,000. Ranged against him were Junot's 25,000, but these were split up into various garrisons around Portugal. His effective moveable troops consisted of just 3,000 men.

Junot's plan to deal with Wellington was simple; he expected an attack either on Loison's division of 6,500 men or Delaborde's of 5,000. Both were little more than 40 to 50 miles from Mondego Bay and each would support one another whilst Junot marched up to deal the decisive blow.

Wellington chose the closest French force, that of Delaborde, who was positioned at Alcobaca.

The Frenchman fell back to better ground around Rolica as Wellington approached. Wellington split his forces into three columns, the left commanded by Ferguson, the right by Trant, whilst he retained the centre. Both Ferguson and Trant were despatched to work around Delaborde, who was waiting on a hill blocking the road to Lisbon.

Seeing the enemy advance, Delaborde threw out pickets and then fell back to a better position on a ridge some 300 feet high, with steep sides. Delaborde intended to stay here until Loison came to his aid.

Wellington regrouped to approach the three gullies which led up the ridge, but the 29th Regiment (under Lake) found itself isolated as the battalions staggered through difficult country. The regiment was the first to come under fire from the ridge and Lake ordered them straight up the gully and into the face of the fire from the French. Lake was killed and the regiment had suffered huge casualties. The French then counterattacked and for two hours Wellington's lead troops could not make any headway.

When Ferguson's men got around the French rear, the pressure slackened and Delaborde began an orderly withdrawal. The French fell back to the Zambugiera Pass, where the British, in hot pursuit, captured half of Delaborde's guns. Still unbroken, they then retired to Cazal de Spraga, where they reformed.

Wellington now broke off his pursuit, angered that he had been unable to finish off Delaborde. As for Loison, he was still at least 10 miles away to the east. Delaborde, meanwhile, fell back towards Lisbon.

Rolica set Wellington on the path to greater things and, as importantly, gave the army a much-needed boost.

Vimeiro, Battle of (1808)

Battle Name	**Dates**	**Campaign**
Vimeiro	21 August 1808	Peninsular War
Troops Involved		
France	13,000 2,100 losses	Commanded by Junot
Britain	17,000 720 losses	Commanded by Wellington

Outcome:
British victory

Just four days after Rolica, Junot marched out of Lisbon to attack the British at Vimeiro. Wellington had been reinforced, but he also had to contend with the political appointment of Sir Harry Burrard as overall commander of the army. Burrard told Wellington that no further offensive actions would be undertaken until Dalrymple and Moore arrived, both of whom were senior to Wellington.

Wellington had deployed his six brigades and eight guns on a ridge parallel to and south of the River Maceira; a battalion guarded the eastern ridge. Two further brigades and six guns protected the village of Vimeiro, atop a flat hill. During the night of 20 August, reports had arrived warning Wellington of Junot's approach, but at dawn there was still no sign of the enemy. At around 0900, the first signs of the approaching Junot were seen heading for Wellington's left flank.

Wellington immediately transferred three brigades from the western ridge to the eastern one and as he watched the French approach, it seemed that the first attack would be made on Vimeiro itself, held by just 900 British infantry. They would face the onslaught of Thomieres 2,400, supported by cavalry and artillery.

As shrapnel cut into the advancing columns, the French soon came within musket range and the first stopped dead as they desperately turned to form line. The second column of 1,200 men fared no better, although it had suffered less casualties in the advance to musket range. With the fleet-footed British infantry firing into the front and sides of the column, it, too, faltered then fell back, leaving seven guns to the enemy.

The French were not finished and launched two more grenadier columns at Vimeiro, once again they were stopped by the 2,000 British now defending the village. One of the columns, however, had worked its way to the south and stormed into Vimeiro; during this time there was intense house-to-house fighting. Despite being outnumbered, the British forced them out and regrouped.

It was now Wellington's opportunity to attack, but he only had barely 250 cavalry under Taylor.

Although they met with initial success, they galloped ahead of the support and suffered fifty per cent casualties, including Taylor himself.

One further attack by Solignac's 3,000 men on the east ridge failed, as did another by Brennier's 3,200. The latter had gained ground but before noon all were fleeing completely disordered and broken.

Corunna, Battle of (1809)

Battle Name	**Dates**	**Campaign**
Corunna	16 January 1809	Peninsular War Danube Campaign
Troops Involved		
France	900 losses	Commanded by Soult
Britain	900 losses	Commanded by Moore
Outcome:		
French victory		

The so-called Corunna campaign had been a near disaster. Moore's retreat through the inhospitable countryside in the height of winter was a nightmare. The Spanish inhabitants were, quite rightly, reluctant to assist as this army of liberation had liberally helped themselves to anything that had not been nailed down. The French, under Soult, were in hot pursuit, eager to eliminate the British army. At times, as the chasing hounds caught up with their quarry, it would turn and fight. For the most part, the British had little fight or resilience left to resist.

On 31 December, Crawford's Light Division split of from the retreat and headed for Vigo, where it was safely picked up by the Royal Navy. Although relieved of the burden of more mouths to feed, Moore was exposed to a much greater extent than ever before.

Between 11 and 15 January 1809, Moore's troops managed to stagger into Corunna, hoping for a quick exit from the Peninsular and escape from the pursuing French. Corunna afforded some protection by the River Mero, but the San Christobal Heights were quickly seized by Soult.

As 16 January dawned, the French launched a massive attack along the whole of the British defensive ring. With a four-to-one advantage in guns, Soult's men desperately tried to throw the British into the sea. The heaviest attacks, supported by artillery, hit the British right flank. Initially, at least, the British gave ground, but stubbornly refused to be beaten. In the centre, the strategic village of Elvina was captured by the French, won back, lost again and regained. Everywhere along the front, the last reserves were thrown in to hold the positions.

At the very height of the engagement, Moore was hit by a canon ball and mortally wounded.

His second-in-command, Baird, was also wounded and by the evening the command had passed to Hope. By this stage, the battle had been reduced to an artillery duel, with both sides exhausted.

The British now seized the opportunity to withdraw to the waiting ships and they were not molested again by the French. When the remnants of the army arrived back in England, in order to avoid alarm, they were disembarked during the night. For three months there would be no British field army on the Peninsular, but soon they would be back. Meanwhile, the only British reserve in all of the peninsula was the 10,000 men in garrisons at Lisbon.

Medellin, Battle of (1809)

Battle Name	**Dates**	**Campaign**
Medellin	29 March 1809	Peninsular War
Troops Involved		
France	16,000 – 18,000 700 losses	Commanded by Victor
Spain	24,000 – 35,000 10,000 losses (many captured)	Commanded by Cuesta

Outcome:
French victory

The infirm and, frankly, incompetent Spanish General Gregorio Cuesta had taken command of the Galician army in July 1808. Bessières had crushed him at Medina del Rio Seco on 14 July, but the surrender of Dupont's 17,500 men had helped to stimulate Spanish resistance. Junot was isolated in Portugal and when he was defeated at Rolica and Vimeiro in August, the Spaniards were beside themselves with the prospect of getting to grips with the French again.

Moore arrived in Spain in September 1808 with some 35,000 men; he found the Spanish armies to be vast but useless. They were little more than a rabble of ill-trained and ill-disciplined opportunists. Then Napoleon took command of the 200,000 French in Spain and Moore was compelled to attack the French for the sake of Anglo-Spanish relations. Ultimately, Moore retreated to Corunna and the British dealt Soult a heavy blow in January 1809. The British evacuated and Napoleon thought that the Spanish campaign was as good as over. When Wellington returned to command the British forces at Lisbon, and train the rapidly-improving Portuguese units, he found the prospect of having to cooperate with the likes of Cuesta more of a liability than of any real value.

As would be seen at Talavera, when the odds were against him, Cuesta chose discretion as the better part of valour, but at Medellin it appeared that he had a ready advantage in terms of manpower and it was here that he impudently chose to attack Victor.

Victor's 18,000 men took up positions to the south of Medellin with the River Guadiana at their backs. The Spaniards, under Cuesta, advanced with confidence, a force of at least 24,000, perhaps as many as 36,000. The French, outnumbered though they were, had more artillery and cavalry than Cuesta.

Cuesta advanced on a four-mile front, hoping to flank the French on both sides. Victor simply retired to a better prepared position and made ready to make a stand. The engagement opened with French cavalry sweeping away the advancing Spanish lancers, then charging home into the distracted and ill-organized Spanish infantry. Within a short period, the whole Spanish force was fleeing and Cuesta nearly fell into French hands.

The mass of routed Spanish troops were easy prey for the French, who seemed to scoop up at least 10,000 prisoners, in addition to those who had fallen in the initial, brief period of fighting.

Eckmühl, Battle of (1809)

Battle Name	**Dates**	**Campaign**
Eckmühl	April 22 1809	Danube Campaign
Troops Involved		
France	50,000 6,000 losses	Commanded by Napoleon
Austria	75,000	Commanded by Archduke Charles
Outcome:		
French victory		

On 16 April Napoleon arrived in Stuttgart to take command of operations in Germany. Berthier had commanded some 176,000 men in the theatre and, as far as Napoleon was concerned, had achieved precious little with them. Napoleon found the situation chaotic, with the Austrians on the offensive and very much holding the initiative. Napoleon seized the opportunity to strike at the over-extended Austrians first, hitting Archduke Charles' centre at Abensberg on 20 April and the next day at Landshut. Davoût was sent in pursuit of what was believed to be an inferior force, but instead, at Eckmühl, he encountered Archduke Charles' main force of some 75,000. At his immediate disposal Davoût had barely 20,000 men; not surprisingly, the Austrians seized the chance to crush this smaller force before more French could move up in support.

Around 40,000 Austrians struck Davoût's left flank, whilst the balance of the force marched around to cut the French off from the River Danbube. By 1300 on April 22nd, Davoût was in dire straits, but help was at hand in the form of Lannes and his 30,000 troops, marching north from Landshut.

Whilst Lannes' troops attacked the Austrian left, Davoût rallied his men to hold the Austrians frontally, giving his fellow general the chance to deploy. More French troops arrived on the battlefield in the shape of Lefebvre, who attacked the Austrian right. This new pressure made the Austrian positions untenable and with his flanks exposed and wavering, Archduke Charles ordered a withdrawal towards Ratisbon on the River Danube.

Fortunately for the Austrians, the French troops were exhausted and unable to pursue. Archduke Charles left 6,000 men at Ratisbon which, on the next day, fell under pressure from the French. The action opened with a cavalry battle, fought outside the walls of the city, but Napoleon was determined to strike at the city immediately and not subject it to a siege.

Lannes' corps assailed the city at 1300 and was repulsed. During the afternoon it was widely believed that Napoleon had been shot, which caused the French to waver, but after showing himself to the army they took new heart, launching a second assault at 1700. Three attempts were made on this occasion to breach the walls and eventually, after losing 1,000 men, the city fell at 1900. By now, Archduke Charles and his army had escaped.

Aspern-Essling, Battle of (1809)

Battle Name	**Dates**	**Campaign**
Aspern-Essling	21-22 May 1809	Danube Campaign
Troops Involved		
France	Day 1: 23,000 – 31,400 Day 2: 62,000 – 70,000 19,000 – 21,000 losses	Commanded by Napoleon
Austria	Day 1: 95,900 Day 2: 90,800 23,400 losses	Commanded by Archduke Charles

Outcome:
Austrian victory

Napoleon had firmly established himself in Vienna, but the problem was that Archduke Charles' army was still intact. The greatest problem, however, was getting at him as the Danube lay between them. The bridges in the area had been destroyed by the Austrians. The opportunity to build a bridge at Nussdorf was lost when the Austrians occupied Stammersdorf and Gerasdorf. The only viable option was Kaiser-Ebersdorf, although this would entail building a series of bridges to the Loban Island and then a further bridge between Aspern and Essling.

The immediate area for the bridge building was secured on 18 May and Loban Island was cleared of Austrian troops the following day. By 20 May, both Molitor's and Lasalle's divisions were across the Danube. Whilst cavalry crossed at 1700 a large boat launched from upstream broke the bridge and it was not until 0300 on 21 May that it was repaired.

Napoleon threw more infantry across the bridge at dawn, but again the structure collapsed. The French, however, occupied strong positions protected by dykes. Aspern was not occupied and when the Austrian infantry appeared at 1430 they made a quick dash for the village. By 1500 it was in Austrian hands; Molitor would have to take it back, which he did, holding it against the growing pressure from Hiller and Bellegarde. The bridge had now been repaired once more and further elements of the French army were making their way across. At 1900 Charles led Hiller and Bellegarde's corps against Aspern, clearing Molitor from the village. Massena counterattacked, taking most of the village back.

During the night, more French troops braved the bridge whilst, with the exception of the Austrians still in parts of Aspen, the whole army had returned. Charles erroneously believed that, come the morning, Napoleon would be gone.

At 0300 the French began their attempt to clear Aspern, achieving this by 0500. Rosenberg then attacked Essling, but Napoleon's infantry struck between the two villages at 0700. The French drove the Austrians back for 2 miles, but their line maintained integrity. All that remained was Davoüt's men to finish them off, but by then the bridge had been destroyed once more.

Without reserves, the French fell back and at 1330 Charles launched his assault. Desperately in need of ammunition, Davoüt was getting a trickle across the river by boats, but it was Bessiere's cavalry that held and counter-charged. Charles then resorted to artillery, but by 1400 the battle was over and by 0500 the following morning Napoleon had returned back across the river.

Wagram, Battle of (1809)

Battle Name	**Dates**	**Campaign**
Wagram	5-6 July 1809	Danube Campaign
Troops Involved		
France	Day 1: 110,000 – 136,500 Day 2: 170,500 32,000 – 34,000 losses	Commanded by Napoleon
Austria	Day 1: 130,000 Day 2: 146,000 40,000 – 43,000 losses	Commanded by Archduke Charles

Outcome:
French victory

After the disasters at Aspern-Essling, Napoleon still held Loban Island; this time he determined not to be hamstrung by a single bridge, but to cross it in force with a secure reinforcement and re-supply route to his rear.

The Austrians had thrown up fieldworks running from the banks of the Danube through Aspern, Essling and Gross-Enzersdorf, but just beyond the town the defences abruptly finished.

On 29 June, a new bridge was built at the old crossing point and a brigade was thrown across to reconnoitre the area. The defence works were thinly manned and prisoners reported that Charles had returned to Pressburg. In fact the Austrians had taken up positions on the higher ground along the River Russbach with elements stretching out through Gerasdorf, Stammersdorf and Langenzerdorf to the Danube.

On the night of 3/4 July, French troops began crossing the Danube, primarily heading out towards Wittan. The crossing was carefully planned for maximum surprise, but Bernadotte's troops were still crossing at daylight and Charles now knew that the French were attempting another attack from the Aspern-Essling area.

Charles had deployed his troops to ensure that he could strike at the French from any direction and cut them off from their bridgeheads. The troops holding the defence works would be expected to hold as long as possible, their inevitable retreat covered by cavalry.

At 2000 on 4 July Outinot had moved to secure the crossing and an hour later French batteries opened up and destroyed an Austrian battery on the Hansel-Ground. The area was secured by 2300 and the French opened up on Gross-Enzersdorf. The old bridgehead was reopened immediately by St Cyr and Molitor, Legrand quickly following. Davoüt crossed at 0600 and took Wittan. Gross-Enzersdorf and Wittan were taken by 1000 and by 1230 the French were established on the opposite bank in force. Elements of Austrian infantry in between Aspern and Essling had been rooted out by French cavalry and were streaming away.

At 1300 Oudinot, Lasalle and Marulaz advanced on Rutzendorf, which they captured at 1400. Massena, meanwhile, cleared the remnants of Austrian opposition from the Aspern-Essling area. By 1800, the French army was ready and poised for an attack on the main Austrian line; the Russbach itself was little more than a stream and would prove to be no real obstacle in itself. Before the Austrians could move to consolidate their positions Napoleon launched an all-out attack. Elements of both Eugene's and Bernadotte's corps broke into Wagram but were ousted. Other elements of Eugene's corps broke through the Austrian lines, but were thrown back when Charles committed his reserves and Oudinot gained a temporary foothold in Baumersdorf. The day's fighting was over temporarily, but hostilities would begin in the early hours of the next day. Napoleon ordered Massena to concentrate around Aderklaa with only Boudet covering the left. Davoüt, meanwhile, was to muster around Grosshofen, Napoleon clearly thinking that Charles would anchor his defence for the next day on Wagram.

The last French attack of the day had convinced Charles to change his plans; he had intended to hold the Wagram to Markgrafneusiedl line, but it now fell to Charles to take the initiative. Charles ordered Klenan to strike from Stammersdorf against Aspern and Kollowrat from Hagenbrünn to hit Leopoldau and Breitenke. He planned that both attacks would get underway at 0100, but neither marched out of their camp until 0400. Charles had also planned for Prochaska's grenadiers to hit Süssenbrünn at 0300, but instead of being at Gerasdorf as had been believed, Prochaska was, in fact, at Sauring. In conjunction with this planned move Bellegarde would link up with Prochaska (of Lichtenstein's corps) by moving to Aderklaa. It would then be Hohenzollern's turn to advance and engage the French (Charles reckoned around 0400). John's corps was then expected to move up to take Leopoldsdorf. If the plan had come to fruition, the French army would have been surrounded and solidly beaten, but as it was, with the delays in relaying orders, it was the moves of Rosenberg that first alerted the French to the danger.

Rosenberg had been ordered, as part of the plan, to attack Grosshofen and Glinzendorf. Massena had been moving on Aderklaa with Nansouty and Arrighi following, Davoüt was moving west when he encountered Rosenberg's troops. Nansouty's artillery and Arrighi's cavalry helped Davoüt force Rosenberg back over the Russbach by around 0600. More bad news arrived at Napoleon's headquarters at Raasdorf around 0900: the whole of the French left was in danger of crumbling.

Bellegarde had occupied Aderklaa and awaited Prochaska's arrival at 0730, then they waited for Koltowrat. It was around this time that Massena arrived to occupy Aderklaa, which Bernadotte should have already occupied. St Cyr seized the village, then lost it again and as Bernadotte's troops moved up, they broke and fled. Napoleon arrived and told Massena to recapture Aderklaa as it was pivotal for the French left. Massena duly obliged when Mollitor's men stormed the village.

Offensive action now rested on Davoüt and he was ready by 0930. Austrian cavalry were dispersed from Siebenbrünn and he set up his artillery here to bombard Markgrafnensiedl and the heights.

On the French left, Boudet had been driven back to Aspern; at least part of Charles' plan was working. Pressure was building elsewhere, Molitor was evicted from Aderklaa at 1100 and the bulk of Massena's corps was heavily engaged to the south of the village.

Meanwhile, back on the French right, Davoüt's troops were storming through the field fortifications, having silenced Rosenberg's artillery. The melee surged through Markgrafneusiedl; now the Austrian rear was exposed and threatened by Grouchy and Montbrun's cavalry. Charles moved quickly to cover with troops from Hohenzollern's command facing Oudinot.

Napoleon now planned to switch the axis of attack; Massena would swing south to cover Klenan and Kollowrat, supporting Boudet and protecting the bridges. Eugene would occupy Massena's positions and hold both Lichtenstein and Bellegarde. Around noon Massena was in place and Davoüt was making steady progress in the west, it was time for MacDonald's troops (of Eugene's corps) to advance.

After a preliminary bombardment from his own eighteen guns, Wrede's twenty-four pieces and Lauriston's guard artillery of sixty guns, MacDonald advanced in a curious open-ended oblong of 8,000 men. The advance took place at 1230; it was first hit by Lichtenstein's cavalry but these were beaten off. Both Bellegrade and Kollowrat threw everything they had at him, but they both failed. MacDonald's men were not halted until 1345, by which time they were over halfway to Süssenbrünn. Davoüt was still pushing Rosenberg back, Oudinot struck at Hohenzollern and snarled his line and linked up with Davoüt.

Klenan began to retire to Stammersdorf and Bellegarde to Hagenbrünn. MacDonald pushed on once more, this time taking Süssenbrünn and Gerasdorf and reaching the outskirts of Jägerhaus, here he was stopped by a reserve brigade of Austrians.

Meanwhile, at 1600 John finally arrived on the battlefield at Siebenbrünn and pushed towards Markgrafneusiedl before Reynier, held in reserve, saw him off.

The casualties on both sides had been vast, probably as a result of the large numbers of artillery pieces on the field. Charles' men were beaten, but many managed to retire in reasonably good order.

Talavera, Battle of (1809)

Battle Name	**Dates**	**Campaign**
Talavera	27-28 July 1809	Peninsular War
Troops Involved		
France	46,000 7,268 losses	Commanded by Joseph Bonaparte
Britain/Portugal/Spain	55,600 (20,600 in combat) 5,365 losses	Commanded by Wellington/Cuesta

Outcome:
British/Portuguese/Spanish victory

Soult had been forced out of Portugal by June 1809, leaving Wellington clear to strike out into Spain with the intention of linking up with the Spaniards. It was agreed that the Allied forces would meet to the north of the Tagus at Plasencia.

The French, meanwhile, had three large forces still in Spain. Joseph Bonaparte was in Madrid with 12,000 men, Sebastiani had 22,000 men at Madridejos and Victor, who was moving up from Merida to Talavera, commanded another 20,000 men.

For Wellington and the Spanish General Cuesta, there was an opportunity to take on each of the French armies and defeat them in detail. Unfortunately, the two men could not agree and, with the Spanish contributing 35,000 of the 55,000-strong combined force, Wellington found it difficult to pin his ageing ally down.

The two men had met on 20 July and missed their opportunity to fall on Victor by the time they agreed tactics on 23 July and Victor was falling back to Madrid on the following day. Cuesta would not be denied and he set off in pursuit of Victor, his troops receiving a bloody nose on 25 July, when the French turned to face them.

Wellington's troops, meanwhile, had reached the River Alberch on 27 July, just to the east of Talavera. Whilst viewing the French from a position at Casa de Salinas, Wellington was nearly captured by enemy light infantry, but managed to make good his escape. The day saw several skirmishes but little more than probes from both of the armies.

During the night Wellington organized his deployment. His line stretched from Segurilla in the north to Talavera in the south. Cuesta's chastened 35,000 were on the right flank (Talavera) and the Allied left flank rested on Cerro de Medellin, a large hill about a mile from the heights of Segurilla.

Shortly after 2200, a French division approached from the Cerro de Cascajal facing the Cerro de Medellin. The French crossed the valley and the Portina, a stream running along the valley. The Cerro de Medellin was occupied by British and German troops who were caught completely unawares by the French. It was a close-run thing as the British tried to beat off the French columns who had reached the summit. Eventually the French gave up, leaving 300 dead, but having inflicted similar casualties on the enemy, as well as having thoroughly destroyed the night's rest.

At 0500 on 28 July the French artillery began firing at the Anglo-Spanish positions. The sixty-or-so guns were answered by British gunners. The French attack consisted of 4,500 men in three

columns, led by Ruffin. They advanced steadily despite stubborn defences. The main force seemed to be heading for Stewart's and Tilson's brigades which they outnumbered. Steady British fire stopped the attacking columns and with little hope of returning effective fire, the assault broke up and soon Ruffin's men were streaming back to the French lines; by 0700 it was over.

The battle descended into a one-hour artillery duel, followed by a five-hour lull. The French artillery opened up once more at 1300. This time the 4,500 men of Laval's division were heading for the Pajar, a fortified farmhouse at the junction of the British and Spanish troops. Campbell's division drove them off, forcing them to leave behind seventeen artillery pieces. No sooner had the attack petered out than 80 French artillery pieces opened fire as 1,500 infantry, under Sebastiani and Lapisse marched forward. Even these veterans could not withstand the British fire power and began to fall back. Some of the British troops were over-zealous in their pursuit and a gap opened up in the line, towards which 22,000 French infantry and cavalry headed. The gap was plugged first by a single battalion and then three more, amounting to 3,000 men. British artillery fire began to plough into the advancing French, then, almost as one, the 3,000 British muskets fired. The French columns were again shattered, with Lapisse alone leaving 1,700 men behind as he fled back across the Portina.

The French now turned their attention to the Medellin itself. Ruffin's division began to advance across the valley under British artillery fire. Wellington threw Anson's cavalry brigade at Ruffin, whose men promptly formed square. A disaster struck the lead cavalry regiment, the 23rd Light Dragoons, who ploughed into a dry river bed and then were attacked by French cavalry; casualties were high. The second regiment, the 1st Light Dragoons (King's German Legion) failed to penetrate the French squares which were still under artillery fire.

Ruffin's men, however, were beginning to waver and they finally pulled back to Cascajal. The fighting petered out and Wellington's troops remained in position, awaiting more attacks in the morning. At dawn Wellington could see that the French had taken enough and had retired.

Busaco, Battle of (1810)

Battle Name	**Dates**	**Campaign**
Busaco	27 September 1810	Peninsular War
Troops Involved		
France	66,000 4,600 losses	Commanded by Massena
Britain	50,000 1,252 losses	Commanded by Wellington

Outcome:
British victory

When Ciudad Rodrigo and Almedia fell into French hands, Wellington was forced to retreat into Portugal, taking up positions on a ridge at Busaco, overlooking the River Mondego. The British positions covered a 9-mile front from Busaco to Mondego. The French could be seen well in advance and the ridge was difficult to climb.

At 0530 on 27 September, French skirmishers began to engage the British pickets, then, emerging out of the fog, came Merle's troops, who hit Lightburne's brigade, but were driven off by volley fire. Four more French battalions tried their luck further south but it, too, failed and seven more under Foy made their way up the ridge. It was 0645 and the fog had been burned away by the early morning sun. It took the intervention of Leith to help fend off Foy's attack and they eventually fell back, Foy himself wounded.

At 0830 the French attacked again. Ney sent Loison's division to tackle Crawford's part of the line which occupied the village of Sula. The French took the village and ran straight into British artillery. Loison refused to be stopped; his men overran the guns and stormed the ridge. Crawford quickly counterattacked with the 43rd and 52nd; Loison's men were routed, leaving over 130 dead. In the attack Loison had lost 1,200 men out of his force of 6,500.

To the south, Marchand led eleven battalions against Pack's Portuguese, but these were roughly handled by the defenders and sent reeling back.

By noon, all of the French attacks had been comprehensively dealt with and the French did not seem to have any more stomach for a determined assault on Wellington's positions.

Massena had seen all of his attempts fail and, what was more, he had lost the highest proportion of officers to men in any action in the Peninsular War. Aside from those captured (which included General Simon of Loison's division), no less than 300 officers had been killed, captured or badly wounded.

Massena was not done; the next day he threw his cavalry around Wellington's flank and the British force retired to Torres Vedras, Massena's next major obstacle.

It had been a bloody encounter and, despite Wellington's previous suspicions about their quality, the Portuguese troops had acquitted themselves well and would remain a solid addition alongside his British regulars for the campaigns to come. Pack's men had come of age in one of Wellington's well-won victories.

Barossa, Battle of (1811)

Battle Name	**Dates**	**Campaign**
Barossa	5 March 1811	Peninsular War
Troops Involved		
France	7,000 2,062 losses	Commanded by Victor
Britain	5,000 1,238 losses	Commanded by Graham

Outcome:
British victory

After the Battle of Busaco, Wellington occupied the lines of Torres Vedras, three defensive barriers protecting Lisbon, a mixture of natural and man-made fortifications. Massena's troops closed on the area, oblivious of their defensive strength until they saw them, instantly realizing that an attack would be pointless and expensive.

Wellington could sit and wait in the certain knowledge that he could not only feed his troops, thanks to the Royal Navy's sea supremacy, but that the British ships would also deliver him reinforcements. He could afford to wait and let Massena's men try to live off the land.

By November 1810, the unequal supply situation had wreaked havoc on the French. When Massena re-crossed into Spain, in April 1811, he had lost 25,000 men. Massena, throughout this period, was leeching reinforcements from Victor, who was laying siege to Cadiz. Soult, meanwhile, had moved to besiege Badajoz. This meant that the 25,000 Allies bottled up in Cadiz were now only facing 15,000 of Victor's men. An opportunity for offensive action against the French now presented itself.

Some 14,000 Anglo-Spanish troops were shipped to Tarifa, where they would attack the French besieging Cadiz from the rear after a march north to the city. Another 4,000 Spaniards from Cadiz would launch a simultaneous sortie out of the city.

Graham's troops were forced to land at Algeciras instead of Tarifa due to poor weather on 23 February, with the Spanish troops arriving on 28 February. From the outset, Spanish General Manuel La Peña, a man not noted for his abilities, claimed command, to which Graham reluctantly conceded.

By 5 March the force had reached Barrosa where it engaged a French force under Villatte. Villatte's men withdrew to deal with the Spanish sortie from Cadiz, which persuaded La Peña that victory was at hand. Graham's men had occupied the Barossa ridge, but La Peña ordered him to join him and launch another attack on the French. Prudently, Graham left a battalion on the ridge. No sooner had Graham moved forward than Leval and Ruffins' divisions began to move on the ridge.

Five Spanish battalions retained on the ridge fled at the sight of the French, so Graham began to order his troops to return to the ridge. Browne, commanding the battalion on the ridge, hit the French first, soon supported by the brigade of Guards, and together they swept Ruffins' men off the ridge. Meanwhile, Whealley's brigade had defeated Leval's division, making the British victory complete without a shot being fired by La Peña's troops.

Albuera, Battle of (1811)

Battle Name	**Dates**	**Campaign**
Albuera	16 May 1811	Peninsular War
Troops Involved		
France	25,000 7,000 losses	Commanded by Soult
Britain/Portugal/Spain	32,000 6,000 losses	Commanded by Beresford

Outcome:
Allied victory

Whilst Wellington was completing his victory at Fuentes de Onoro on 5 May 1811, around 130 miles south, Beresford was beginning to lay siege to Badajoz. No sooner had he begun his operations than he heard that Soult was on the march from Seville with 25,000 men, intent on driving Beresford off.

By 13 May, Beresford was marching south to intercept Soult and on 15 May he reached Albuera. He positioned his troops with the town anchoring his centre and the other troops occupying heights to the west of the town. The French arrived from the southeast on the morning of 16 May and drew up their lines facing west.

The first attacks hit Albuera itself, held by Alten; more probes were made to the south of Albuera Bridge. The real threat faced the Spaniards holding the right of the Allied line. Soult had discovered that the River Albuera was fordable in several places and had crossed with considerable infantry and cavalry, shielded by a heavily-wooded area. The Spaniards, under Blake, were ordered to meet the French with the promise of British regulars in support. Beresford was sure that Soult's main attack would come from the east, but Soult was about to throw 8,400 men at Blake. Supported by the British, however, the Spaniards stopped the French columns, but then it began to rain.

Out of the smoke and gloom rode the French cavalry. Colborne's brigade, which had inflicted heavy casualties on the French infantry, were cut to pieces; only 300 of 1,600 survived. The French swept on, Beresford checked the 7,800 French with two fresh British brigades (3,700). A vicious and ruinous firefight lasted for over an hour, but the French numbers began to tell.

Around midday, Beresford ordered Carlos de Espana's Spanish brigade forward, but they refused. Instead, and against orders, Lowry Cote advanced to support with 4,000 fresh British and Portuguese troops. This turned the tide. He beat off repeated French cavalry charges and fell into line with Stewart's remaining men. The French began to waver under the fire, exhausted and now under pressure.

The French streamed to the rear, but the heavy rain denied the British the opportunity to pursue. After four hours of heavy fighting, Beresford's men had grabbed victory from the jaws of defeat. Even Soult realized that the steady British infantry had refused to recognize that they had been beaten. It was a Pyrrhic victory, with 6,000 dead and wounded.

Badajoz, Battle of (1812)

Battle Name	**Dates**	**Campaign**
Badajoz	16 March – 6 April 1812	Peninsular War
Troops Involved		
France	5,000 5,000 killed, wounded or captured	Commanded by Phillipon
Britain	25,000 5,000 losses	Commanded by Wellington

Outcome:
British victory

Badajoz was the strongest French fortress on the Portuguese border, with nine bastions, walls up to 46 feet high and a castle within its precincts. On the right bank of the River Guadiana there were three impressive outworks and on the left bank there were three more. Having captured Ciudad Rodrigo, Wellington now fixed on taking the fortification, deciding to launch his first attacks on the south of the city.

On 16 March his men began to dig in, the French launching a sortie three days later. Each side prepared for the inevitable assault, the French even creating a dam to make a lake which would force the British to march across the front of their main batteries.

Considerable progress was made on 25 March when the British took Port Picurina on the left bank and it was immediately occupied by British guns. News that two French armies were on their way to relieve the garrison forced Wellington's hand; he knew that he must launch his attack soon or all of his efforts would be for nought.

Wellington gave orders for the assault to go in on the night of 5 April, but the breaches in the Santa Maria and Trinidad bastions were insufficient for success, so he postponed for twenty-four hours. During the day of 6 April, the French feverishly shored up the breaches and prepared.

The British, too, were ready and at 2140 the attack was underway. Wilson's troops took San Roque on the left bank, then the storming parties headed for the breaches. The French had placed mines, set up killing zones and placed sharpened stakes to prevent the British from getting through. Despite making forty attacks on the breaches, not a single British soldier got into the town and 2,000 had been killed or wounded.

Meanwhile, Picton had been trying to storm the castle itself, crossing around the dam to approach the walls. He, too, had been denied, but Colonel Ridge managed to get some of his men on to the ramparts. More British troops poured over the walls and soon the castle was in British hands.

Phillipon had relied on the castle holding and had intended to use it as a last line of defence should the British carry the breaches. He and his men were now faced with disaster and by 0200 on 7 April resistance had collapsed and Badajoz was in Wellington's hands.

Salamanca, Battle of (1812)

Battle Name	**Dates**	**Campaign**
Salamanca	22 July 1812	Peninsular War
Troops Involved		
France	50,000 14,000 losses	Commanded by Marmont
Britain	48,000 – 48,600 5,214 losses	Commanded by Wellington

Outcome:
British/Portuguese victory

On 13 June Wellington advanced from Ciudad Rodrigo and marched into Salamanca four days later. The three small French-held forts were besieged and fell on 27 June. The French army closed in on the area and remained close at hand.

Bizarrely, on 18 July, both armies marched, shadowing one another either side of the River Guarena. On 21 July, the two armies crossed the River Tormes and set up camp within a few

hundred yards of each other and the battle now seemed inevitable.

The following morning saw the two armies heading south in parallel; it was a race to either secure the supply routes from Ciudad Rodrigo as far as Wellington was concerned, or to sever them, which was Marmont's plan. French troops collided with the leading Portuguese brigade at Los Arapiles; the Portuguese were driven off, leaving the French in possession of a 100-foot-high hill, the Greater Arapil, to the northeast of the village.

The bulk of the two armies swung southwest. Marmont, from the Greater Arapil, wrongly assumed that Wellington was retreating back to Ciudad Rodrigo, but the dust he saw in the distance was, in fact, Pakenham's 3rd Division coming up to Aldea Tejada to protect Wellington's right flank.

By the early afternoon, most of Wellington's troops had stopped near the Lesser Arapil, half a mile north of Marmont's vantage point. By now, intent on cutting off Wellington, the French were strung out over 4 miles, with the divisions under Thomieres, Maucune and Clausel in the lead. Wellington was having his lunch at Los Arapiles, peering at the French through his telescope. Seeing how stretched the French army was, he realized that this was his opportunity. Instantly, he remounted and galloped off to Pakenham with instructions that he should attack at the earliest opportunity.

Around 1530, Pakenham's troops, supported by a thousand cavalry, appeared on Thomiere's right flank. The 6,000 British and Portuguese poured heavy rocket-fire into the dense and confined French. Thomiere himself and the two leading French battalions were wiped out, his now-leaderless division soon evaporating.

Just after 1700, Bradford's Portuguese (8,500 men) hit Maucune's division, which had been following Thomiere. Maucune, having seen the British attack on Thomiere and expecting a cavalry presence, ordered his 5,000 men into square. Unfortunately, it was the wrong decision - although his artillery wreaked havoc on the Allies, Maucune's squares were decimated by three well-controlled and devastating volleys from the infantry. As the French ran, it was then that the Allied cavalry were unleashed; nearly five battalions of Maucune's men were slaughtered.

Brennier had been quickly moving up his 4,300 men and as he saw the onrushing dragoons, he, too, formed square. Either his men were too exhausted or simply scared, but the volleys did nothing to break the charge. Brennier's men, too, were swept aside and were running.

The Allied dragoons, under Le Marchant, were completely out of control, pursuing the French in every direction. In forty minutes, three French divisions had been accounted for but, tragically, Le Marchant was killed during the pursuit.

To the east of Los Arapiles things were not going so well for Wellington. To the east of the Greater Arapil, Cole's 4th Division had been beaten off by two fresh French divisions. At the Greater Arapil itself, Portuguese troops, under Pack, had also been flung back by the French.

Marmont had been badly wounded and his next in line, Bonnet, too; overall command now lay with Clausel, who still firmly believed that a victory could be won. He ordered Sarrut's division to hold Pakenham whilst he launched 12,000 men across the plain between the Greater and Lesser Arapils. The manoeuvre was doomed; this was exactly what Wellington had expected.

Clinton's 6th Division held the centre, with Anson's brigade (of Cole's 4th Division) on the left and Portuguese troops under Spry on the right. The whole line was reinforced by more Portuguese under Rezende. In the face of Allied artillery fire from the Lesser Arapil, the French marched on in column. When they reached musket range the heads of the columns were blown away with carefully-aimed volleys. The French tried to return fire, but the volleys were too effective and they were driven back, shattered and depleted.

By sunset, Clinton's troops had taken the Greater Arapil but ran into 5,500 fresh French men under Ferrey. Unlike his compatriots, Ferrey deployed well and drove Clinton back, only being forced to retreat himself when his flanks were exposed.

The whole of the French army should have been lost, the Alba de Tormes Bridge was supposed to be guarded by Spanish troops, but these men, under Carlos de Espana, had withdrawn. Hundreds of French were taken, along with twenty guns, and on 12 August Wellington marched triumphantly into Madrid, which had been occupied by the French since December 1808.

Mogilev, Battle of (1812)

Battle Name	**Dates**	**Campaign**
Mogilev or Saltanouka	23 July 1812	Russian Campaign
Troops Involved		
France	26,000 1,000 losses	Commanded by Davoüt
Russia	49,000 4,000 losses	Commanded by Bagration

Outcome:
French victory

The 2nd Russian Army was steadily moving east, pursued by Poniatowsky, Latour-Manbourg and Jérome Bonaparte. Further to the east was Davoüt and it was now time for a vast turning movement to trap the Russians and stop their eastward flight. Davoüt headed for Mogilev to close the door and entered the city on 20 July. It remained imperative for the Russians that Bagration's troops should be able to cross the Dneiper. Raevski was at Stary Bykhov and headed for Mogilev to support Bagration. Just outside the city, his advanced guard cavalry had a sharp fight with French horsemen, which he managed to overwhelm.

This new threat alarmed Davoüt and he determined that he would have to fight to hang on to Mogilev after all. He despatched up to 28,000 men to the nearby village of Saltanovka and prepared for the attack. Meanwhile, Bagration was making all speed to Mogilev and by the morning of 23 July his advanced guard reached the edge of the forests to the south of Saltanovka. Initially, Dessaix engaged Paskevich, each trying to turn one another's flank in the thick woodlands. Raevski had established himself on the right wing and proposed to attack as soon as all of his troops had moved up. He then received a despatch from Paskevich telling him of his failure to turn Dessaix and the fact that Davoüt was said to have five divisions (two of cavalry) at his immediate disposal. It was now evident that Raevski was not strong enough to handle the French at Mogilev. Whilst Raevski held as long as he could, Bagration made for the Dneiper, constantly harassed by the pursuing French. Reluctantly, Raevski returned and headed for Novoselky and then to Stary Bykhov.

By 23 July, all significant forces were over the Dneiper and the Russian armies reunited at Smolensk on 3 August.

Davout had held firm at Mogilev when there was every chance that Raevski could have broken through and linked up with the Russian 1st Army under Barclay; as it was he had bought Bagration the time he needed to escape the trap set for him by the French.

Of particular note at Mogilev are two interesting facts; firstly Bagration (and presumably Raevski) thought that they were facing at least 60,000 French, which implies Davoüt's powers of manoeuvre were particularly strong. Especially interesting was the presence of Raevski's sons, the elder claimed the flag of a Smolensk regiment and the younger was always at his father's side.

Borodino, Battle of (1812)

Battle Name	**Dates**	**Campaign**
Borodino	7 September 1812	Russian Campaign
Troops Involved		
France	133,000 30,000 losses	Commanded by Napoleon
Britain	120,000 44,000 losses	Commanded by Kutusov

Outcome:
French victory

From the mid-afternoon of 5 September, the Russian army began occupying and preparing field fortifications around Borodino. The new overall Russian commander was Kutusov, who realized that in a war of manoeuvre, strategy and tactics, he had little hope of beating Napoleon. Nonetheless, he was charged with protecting Moscow from the French. First he needed to postpone the inevitable battle and then be prepared to meet Napoleon on ground that favoured defence.

The Borodino site was well chosen. To begin with it was on higher ground. The town itself was strongly held, although in advance of the main line. The left flank rested on Utitza then ran through woodlands to the Bagration Fléches (a series of fléches or redoubts), south of Semyonovskaya. Beyond the village was the Great Redoubt, the pivotal point of the line. The Russian defences then covered the higher ground above Borodino, continuing past Gorki behind the Kalatsha River to its junction with the Moskvo River.

To crack the impressive defence works, Napoleon decided to try to force the Russians into a cul-de-sac between the Kalatsha and Moskvo Rivers. Consequently Davoüt would move first against the fléches and Poniatowski would press up the Old Smolensk Road and take Utitza. Ney was to attack Semyonovskaya, Eugene would then take Borodino and move on the Great Redoubt. All Kutusov could hope for was that his men would stubbornly hold their ground as he had little in the way of reserves to commit should the need arise.

At 0600 on 7 September the French attacks got underway with a bombardment of the Russian left. Ney's artillery began to strike at the enemy centre and Eugene began to soften up the Russians in Borodino.

Davoüt's troops got underway shortly after Poniatowski started to move. Unfortunately, Poniatowski chose to cut across country to the Old Smolensk Road, the net result was that his troops did not get into action until around 0800. By this time, Eugene had launched Delzons at Borodino.

Bagration's troops held the Russian left and he sent Tutchkov to help hold off Davoüt. It met with success and more Russian infantry were committed, then cavalry; Davoüt was stunned by a shell and Dessaix badly wounded.

At Borodino, Delzons had forced out the Russian Guard and began to advance towards Gorki; here he was met by thousands of Russians and driven back with high casualties. The Russians then launched a counterattack to take back Borodino, but they only managed to destroy the bridge to the east of the town.

Morand was leading the assault on the Great Redoubt and had, so far, driven in the light infantry covering the fortification. At 0700 Ney moved to attack and Junot and Murat advanced to offer closer support. Bagration began to panic, seeing the fresh French troops, and pulled Raevski's reserves to the front. Meanwhile, the long process of getting Bagavout's troops from the far right wing had begun.

Shortly after 0800 Ney's troops stormed two of the three fléches, the third also being overrun after it was abandoned. Bagration launched a counterattack with cavalry and took back all-but the northern fléche which was now held by Württemberg's infantry under the temporary command of Murat, who had ridden forward to reconnoitre the area. He was supported by Nansonty, who in turn supported Ney's renewed attacks. By 1000, the whole of the ridge below Semyonovskaya was in French hands.

Meanwhile, Poniatowski had eventually taken Utitza and was heavily engaged in the surrounding wooded area. At 1000 Eugene made an attempt on the Great Redoubt, but this was repulsed. After Bagration's counterattack, Junot was pushed into the line between Ney and Davoüt, he was then ordered further south to link up with Poniatowski to help push the Russians out of the woodlands.

Platov and Uvarov's cavalry had not yet been committed and these were ordered to cross the Kalatsha River and fall on Eugene's left flank; they would get into action some time after 1200. The fléches were taken back by another of Bagration's counterattacks, but Ney and Davoüt had regained them by 1130. The Russian left now held a line on the Semoyonovskaya Creek, where they rallied and held off attempts to take the village. At around this time, Bagration was mortally wounded and fresh French troops were moving up to support the attacks on the Semoyonovskaya area. Napoleon despatched Friant's division to reinforce Ney as Murat threw Nansonty and Latour-Manbourg across the creek to open up the opportunity for Friant to storm the village. Nansonty suffered heavy casualties and was forced to retire, but Latour-Manbourg cleared the

Russian infantry and Friant duly took the village. The heights were now in French hands and from here their artillery could do much damage to the massed Russians.

By 1200 Tutchkov had pushed Pomatovski back to Utitza and was holding off Junot, but, as the Russian centre was pushed back, both Junot and Pomatovski resumed their attacks. This time it succeeded and the Russians fell back on Old Smolensk Road.

Meanwhile, in the Great Redoubt area, the position was finally taken by Morand at 1100. Desperately, Russian artillery was trained on the French, followed by a counterattack by Raevski, which snatched back the redoubt.

Uvarov and Platov now made their presence felt on the French left, initially causing much panic. Delzon's infantry formed squares to block Uvarov and French baggage guards held off Platov. Nonetheless, it carried enough consternation to draw in Grouchy and Lecchi, giving Kutusov time to reorganize his lines.

Ostermann's troops reinforced Raevski, Bagavout now held the Russian left and Docturov took the right, resting his flank on Gorki. The redeployment was made just in time as Eugene redoubled his efforts to throw the Russians out of the Great Redoubt. His infantry forced out Raevski's troops and Caulaincourt (who had replaced the dead Montbrun) scattered the Russian cavalry that had been committed. His command was brief as he, too, fell near the Great Redoubt.

Ostermann now counterattacked but was roughly handled and fell back. Grouchy then swept forward to mop up the fleeing Russian infantry, but was stopped by the Russian Guard cavalry.

The Russians now occupied the next ridge stretching from Gorki down to the Old Smolensk Road; only a massive and coordinated attack would shift them. The French Guard had not been committed, but Napoleon decided not to risk his last reserves. The battle now rested on artillery fire, a solid pounding from the French would dissuade any attacks from the Russians.

Napoleon fully expected to resume in the morning, but after having written to the Tzar, proclaiming a great victory, Kutusov ordered a retreat in the early hours of 8 September. Kutusov had promised to defend Moscow but Napoleon pursued him vigorously, leaving Junot on the battlefield at Borodino. Murat caught Kutosov on 9 September and captured 10,000 wounded men. Pausing to the west of Moscow on 13 September, it was quickly realized that the city was indefensible and Kutusov began to retire towards Koloman as Poniatowski had already taken Kaluga. Murat reached the gates of Moscow on 14 September.

Lützen, Battle of (1813)

Battle Name	**Dates**	**Campaign**
Lützen	2 May 1813	Leipzig Campaign
Troops Involved		
France	45,000 – 110,000 20,000 losses	Commanded by Napoleon
Russia/Prussia	73,000 18,000 losses	Commanded by Wittengstein/Blücher
Outcome:		
French victory		

Napoleon was on the move and the Allies began concentrating between Leipzig and Altenburg in preparation to attack the French right flank near Lützen if the French continued to advance over the River Saale. From 1 May, the French advance came under increasing pressure and, at Weissenfels, Bessières was killed in a heavy engagement.

As the French began to advance towards Lützen, the Allies planned to move to engage at 0100 and complete their manoeuvres by 0700 on 2 May. This was optimistic and by 1100 they had reached Gross-Görschen, where they could only see scattered groups of French.

Wittengstein was confident of victory and ordered Blücher to attack at 1145. Instead of encountering a small force, they found themselves up against two divisions; the French occupied Gross-Görschen and Starsiedel to the west and Marmont was already on his way to support.

By 1200, Southam's troops in Gross-Görschen had been driven out by artillery, but Girard, at Starsiedel, held. Ney now arrived on the scene and organized a counterattack to stop Southam's retreat and a general action got underway, the sound of which attracted Napoleon's attention. He was deeply concerned about the developments on his right. He needed to buy time to allow Marmont to move up, whilst MacDonald swung his corps southwards to attack the enemy right over the River Flossgraben. The Guard would move up to Kaja to deliver the main infantry attack.

When Napoleon reached the battlefield at 1430, Ney's men were exhausted and in dire straits. Bertrand's advance on the enemy left had been halted by the arrival of Russian troops and all along the front the French were under heavy pressure. Finally, the Guard arrived at Kaja, but Napoleon did not commit them immediately. All that was required was to hold, MacDonald was coming up on the enemy right and Bernard had resumed his advance on their left.

Blücher was wounded and Yorck took over. Russian reserves were slow in advancing, so Wittengstein decided not to rush Yorck's men to support Blücher's exhausted troops. Finally, at 1600, the reserves had arrived and Yorck charged Kaja to be stopped by the Young Guard.

At 1730 MacDonald stormed Eisdorf and Bertrand linked up with Marmont and at 1800 the whole French army attacked. The French swept through Rhana, Klein-Görschen and Gross-Görschen and the Allies began to withdraw. In the growing darkness the French continued, Marmont being marked by Prussian cavalry at 2100; it had been a costly victory.

Vittoria, Battle of (1813)

Battle Name	**Dates**	**Campaign**
Vittoria	21 June 1813	Peninsular War
Troops Involved		
France	66,000 8,000 losses	Commanded by Joseph Bonaparte
Britain	80,000 5,100 losses	Commanded by Wellington

Outcome:
British victory

Wellington was keen to convince the French that his 1813 campaigns would be aimed at central Spain, when, in fact, he intended to march through northern Portugal and aim for Burgos, then the Pyrenees and into southern France. There were enormous benefits to this as he would be able to open up the northern Spanish supply ports and not have to rely on Lisbon as his main source of materiel.

By 3 June, Wellington's army had reached the Douro, catching the French entirely by surprise.

Burgos was abandoned on 13 June and by 19 June Wellington was to the east of Vittoria. Joseph was hurrying to intercept, followed by a second force, under Clausel (14,000 men), who would not arrive in time to fight. On the morning of 21 June, the French line was in position, with the commanders sitting on Arinez hill at their centre.

Wellington sent Hill's 20,000 men to take the south of the valley, which he achieved at 0830. Two hours later, Graham's troops (British, Portuguese and Spanish) probed west and became embroiled in stiff fighting for several hours on the north bank of the Zadotra River.

Just after noon, Kempt's brigade seized the Tres Puentes Bridge over the Zadorra. He was swiftly supported by the rest of the Light Division, who stormed the bridge at Villodas, whilst Picton took the Mendoza Bridge. Together, they now swept the French off Arinez.

By 1500, the British were pressurising the French at the village of Margarita (on the right flank of the French line). French infantry, under Gazan, supported by artillery, held the south of Arinez but Margarita fell and the French line was compromised.

With D'Erlon being pulled back by Picton, Kempt and Dalhousie, Hill's troops surging from Puebla and Graham's troops closing the net from behind, the French were in a perilous position.

At 1700, Cole's 4th Division snarled in Gazan's troops and the gap separated D'Erlon, who was still being pushed back. As the time passed, Gazan realized he would be cut off. Joseph was viewing all of this with increasing alarm and finally ordered a general withdrawal.

Instead of a measured retreat, the French army disintegrated and the losses totalled, with 151 or 153 guns captured. Joseph's army was saved from complete capture as the Allies fell on the accumulated French plunder. It is said that no greater amount of treasure has ever fallen into enemy hands.

Dresden, Battle of (1813)

Battle Name	**Dates**	**Campaign**
Dresden	26-27 August 1813	Leipzig Campaign
Troops Involved		
France	Day 1: 35,000 – 70,000 Day 2: 150,000 10,000 losses	Commanded by Napoleon
Russia/Prussia	Day 1: 120,000 Day 2: 170,000 38,000 losses	Commanded by Schwarzenburg

Outcome:
French victory

The French commander of the city of Dresden, St Cyr, had improvised defences around the suburbs and created five redoubts. The city's two bridges would have to be held at all costs, to provide reinforcements and supply as well as an escape route if needed. The Allies approached the city at around 0500 on 26 August. Wittgenstein would cover the area closest to the River Elbe, whilst Prussian units would take the Great Garden on the outskirts of the city. As it was the Prussians did not take the Great Garden until 0730, Wittgenstein coming under heavy French artillery fire from the east of the river.

At around 0900 combined Prussian and Russian pressure had succeeded in taking nearly all of the Great Garden, whilst the Austrians to the northwest stormed the villages of Plaüen and Löbtans. It was 1100 and St Cyr's men had only been pushed back to their main defensive positions, but now Napoleon arrived and it was hotly debated whether the Allies should retire. As it was, the main Allied attack had been planned for 1600; against the intentions of the Allied commanders, it went ahead.

Wittgenstein was stopped in his tracks by French artillery; the Prussians took the Great Garden, but faltered before the city's walls. The Austrians, however, managed to storm some of the outlying redoubts and break into the suburbs of the city.

At 1730, Napoleon launched his counterattacks, with all of St Cyr's men committed to the defence. Mortier took back the Great Garden and swept Wittgenstein aside and Ney struck at the Allied centre. There was feverish work to be done overnight.

At 0600 on 27 August Nansouty, Ney and Mortier struck the Allied left, clearing the Blasewitz Woods, Seidnitz, Gruhna and Strehlen. St Cyr moved up to establish his line at the latter village.

Murat's attack had been launched at 0630; he quickly overran the village the Austrians had taken the day before and pinned a large enemy force against the River Weisseritz at Dölzschess. The vital Pläven Bridge had fallen and the men were trapped; by 1500 the Allied left was shattered.

On the French left, Napoleon ordered an assault on Reick which was taken at the second attempt by bayonet with Wittgenstein's men returning to Torna. St Cyr was repulsed three times in his attacks against Leubnitz. Whilst Napoleon was planning a new assault on the Allied positions, they decided to retreat as they were short on food and ammunition, but, above all, because their morale was at breaking point.

Leipzig, Battle of (1813)

Battle Name	**Dates**	**Campaign**
Leipzig	16-19 October 1813	Leipzig Campaign
Troops Involved		
France	Day 1: 170,080 Days 3/4: 195,000 73,000 losses	Commanded by Napoleon
Allied	Day 1: 257,000 Days 3/4: 365,000 52,000 – 54,000 losses	Commanded by Blücher / Schwartzenberg / Bennigsen / Bernadotte

Outcome:
Allied victory

The French botched their opportunity to crush the shattered Allies after Dresden and only Vandamme's corps pushed them with any real vigour. Vandamme tried to cut off the Austrian supply and communications at Kulm on 29-30 August 1813, but ran into a large Austrian force. The outcome was inevitable and only part of his corps managed to escape; he was captured along with 13,000 of his men. This disaster was a double blow for Napoleon: Vandamme had been an able and aggressive general and would be sorely missed; equally it gave the Allies fresh hope and helped eradicate the memory of the defeat at Dresden.

There were to be more disasters; at Dennewitz Ney collided with a Swedish force under Bernadotte and was severely mauled on 6 September. Napoleon was now determined to concentrate his forces, choosing Leipzig on the River Elster as his preferred base of operations. Again things did not go well for the French. Blücher and Bernadotte were moving on the city from the north and the Austrians, commanded by Schwarzenberg, were approaching from the south. Napoleon's Confederation of the Rhine was now under dire threat of being crushed. This was not helped by Bavaria switching sides in early October and declaring war on the French.

Some 122,000 men had been gathered at Leipzig by Napoleon by 15 October, he hoped that he could smash each successive Allied army that dared to assail his forces. Leipzig was destined to be the largest battle of the era and would become known as the 'Battle of the Nations'.

The engagement opened on 16 October when Barclay de Tolly's 78,000 Austrians struck against the south of Leipzig. The attack was poorly coordinated and the French were able to hold their counterattack. Meanwhile, in the north, the Prussians under Blücher attacked Marmont's positions around Mockern. Although Blücher enjoyed a two-to-one advantage, he could make little headway.

The Austrians launched another assault to the west of the city before the end of the day, this time the front was held heroically by Bertrand's troops. Although all of the assaults had ended in failure, what was clear was that the city was almost totally surrounded. Equally, the Allies could confidently expect additional reinforcements to move up during the night; Napoleon would have to face even greater odds in the coming days. For once, the initiative lay with the Allies. Napoleon could not dare to concentrate a strong strike force to assail one of the Allies and knock them out of the conflict for fear that one of the other armies would exploit the weaknesses made in his defences by the concentration.

On 17 October, there was little real fighting along the perimeter; the Allies were redeploying their forces and welcoming the arrival of Bernadotte's Swedes and the Russians under Bennigsen, amounting to a further 100,000 men. The ring around the French now consisted of Austrians and Russian to the south, Swedes in the east and Prussians in the north, but they had not yet closed the route to the west. Despite this lifeline to the west, the French were in great need of ammunition; each day of conflict depleted their low reserves even more and this would be a telling factor as the engagement developed.

On 18 October, the Allies were finally redeployed and Napoleon had spent the previous day shoring up his defences and the perimeter. The assault called for the coordinated attacks of some 350,000 men. The pressure along the whole of the perimeter was intense, but the breaking point came when the Saxons, hitherto fighting with Napoleon, defected to the Allies. Napoleon's only hope now was to extricate as much of his army out of the perimeter as possible and conduct a fighting withdrawal to the west. For nine hours, the Allies battered the French, eventually penetrating the city itself. It was only then that Napoleon ordered a retreat. As if the position was not perilous enough, the French conspired to do themselves an even greater injury. With still some 20,000 French manning the rearguard on the Leipzig side (east) of the River Elster, the only bridge was demolished by French engineers. Also trapped on the wrong side of the river were Lauriston and Reynier (who were both captured) and MacDonald and Poniatowski (the latter drowned whilst swimming across the Elster). The French army, although badly mauled, headed southwest to Frankfurt-am-Main and on 30-31 October, their former Bavarian Allies, under

Wrede, attempted to stop them at Hanau. After bombarding the Bavarians with cannon fire, Napoleon's exhausted men cut through their lines to escape.

Napoleon lost another 6,000 men at Hanau and a further 10,000 in the crippling retreat from Leipzig. Of the 170,000 men at Leipzig, 100,000 had been lost. More perilous still, the Allies were now advancing unchecked towards France; Germany had been lost forever.

Brienne, Battle of (1814)

Battle Name	**Dates**	**Campaign**
Brienne	29 January 1814	Campaign in France
Troops Involved		
France	30,000 3,000 losses	Commanded by Napoleon
Prussia	20,000 – 25,000 4,000 losses	Commanded by Blücher

Outcome:
French victory

Napoleon had decided to try to deal with the Prussians first. The most promising target was Blücher's Army of Silesia as it was closer and it was scattered. Napoleon knew that Blücher was at St Dizier, Yorck was far behind and Marmont was already between the two forces. He ordered Marmont to Bar-le-Duc to ensure that no junction between Blücher and Yorck could be made. Mortier and Gérard would be assigned to delaying Schwarzenberg.

Victor had been trying to hold Blücher at St Dizier, but he was gradually being forced back towards Vitry. Napoleon hoped to catch the Prussian with half of his troops separated by the River Marne. By the time that Napoleon had moved up, Blücher had already passed through St Dizier, where he left a rearguard, and was heading for Brienne. The French retook St Dizier, but it was too late; if there was to be a battle, it would be at Brienne.

Napoleon marched towards Brienne in three columns, knowing that Blücher could now link up with Schwarzenberg any day. The French needed to destroy Blücher before he could be reinforced. Napoleon wrote a despatch to Mortier, telling him to move to Arcis-sur-Aube but, unfortunately for him, this instruction was captured by Cossacks and by early 29 January Blücher was aware of Napoleon's intentions.

Brienne was held by Olssufiev's Russians and Blücher now ordered Sacken to march to Brienne from Lesmont and Pahlen's 3,000 Austrian cavalry to head for the town. Blücher had reversed his march to Arcis-sur-Aube just in time; he would be at Brienne to meet Napoleon.

Grouchy's cavalry and artillery began the engagement in the late morning of 29 January and by 1500 Ney and Victor had arrived and were thrown straight into the battle. Ney advanced with two divisions against Brienne, whilst Victor tried to storm the chateau to the west of the town.

With the rest of the French II Corps heading towards Bar-sur-Abe in an attempt to cut Blücher off, the French fortunes on the field were mixed. Throughout the afternoon, Pahlen was held off successfully, but Victor was forced to give ground. Even Napoleon narrowly missed falling into Cossack hands at one point.

The fighting continued during the evening and by now Victor had succeeded in taking the chateau, almost capturing Blücher and Gneisenau. The last shots were fired at around 2200, ending with an abortive Allied attempt to seize the chateau back.

By 2300, Blücher was retiring towards the south; his army was intact and if anything the French had only succeeded in making his junction with Schwarzenberg more certain.

La Rothière, Battle of (1814)

Battle Name	**Dates**	**Campaign**
La Rothière	1 February 1814	Peninsular War Campaign in France
Troops Involved		
France	30,000 – 40,000 6,000 losses	Commanded by Napoleon
Prussia	53,000 – 110,000 6,000 losses	Commanded by Blücher

Outcome:
Prussian (Allied) victory

Following his defeat at Brienne, Blücher managed to link up with Schwarzenberg and determined to launch another attack on Napoleon. He had a formidable force at his disposal: his own two corps and two of Schwarzenberg's gave him 53,000 men. The 25,000 Bavarians under Wrede would tackle the French left and he had been promised the immediate support of Barclay de Tolly's Russian reserve of 33,000 men.

On the morning of 1 February, the vast Allied armies marched north from Trannes to engage. At 1000, Napoleon ordered Marmont to occupy Lesmont and Ney to move up through Brienne. At around 1200 Victor reported large-scale enemy troop movements between Trannes and Éclance. Napoleon had assumed that the main target was Troyes, but he had been mistaken, although he was still unconvinced that this was Blücher's main attack.

Marmont took up positions near Morvilliers, Victor held the centre between Chaumesnil and Le Petit Mesnil (with his right at La Rothière), whilst Gérard occupied the area around Dienville. Nansouty and Grouchy were held in reserve behind the centre and further back (to the southeast of Brienne) was the Young Guard.

At 1300, Blücher launched a major assault on La Rothière. The French cavalry beat off the attack and Guyot's horsemen even managed to overrun a Russian artillery battery. The Russians themselves succeeded in capturing twenty-four French guns in the struggle around La Rothière, the pendulum swinging this way and that as fresh troops poured in.

In the midst of a blizzard, the battle raged on, but the Allies could make no headway against Dienville and Duhesme and still clung on to the northern end of La Rothière. At 1600, Wrede's Bavarians struck at Chaumesnil and Barclay's Russians were driving the French out of La Rothière. Ney counterattacked at La Rothière whilst the Young Guard bolstered Marmont's flagging efforts against Wrede.

By now, with deep snow and exhaustion setting in, the odds were simply too high for even Napoleon. It took all of Napoleon's skills to extricate his men and the retreat was a moral and material disaster. For two days, thousands of French fell due to exhaustion, so intent was Napoleon to link up with Mortier.

La Rothière had been a disaster; honours were even on the battlefield, but Napoleon had allowed himself to face a foe nearly three times his strength and ultimately his men had paid the price.

Craonne, Battle of (1814)

Battle Name	**Dates**	**Campaign**
Craonne	7 March 1814	Campaign in France
Troops Involved		
France	37,000 – 40,000 5,400 losses	Commanded by Napoleon
Prussia	85,000 – 90,000 5,000 losses	Commanded by Blücher

Outcome:
French victory

By 5 March, the French line had retreated to the west bank of the River Seine and Schwarzenberg was in control of Troyes and all of the crossings on the Upper Seine. Paris was under immediate threat and Napoleon surged forward to engage Blücher at the earliest opportunity.

On 6 March, his troops advanced from Berry, having captured the bridge there with a dashing charge by his Polish lancers. Napoleon then received the news that the Plateau of Craonne was teeming with Prussians and he wrongly assumed that this was Blücher's rearguard. Consequently, Ney was ordered to engage, in doing so postponing his attack on Lâon. Napoleon hoped that this would give Marmont and Mortier time to overcome the Prussian garrison at Soissons and join him on his offensive.

Blücher was, in fact, calling Napoleon's bluff and wanted him to engage at Craonne. He proposed an elaborate trap to lure the Frenchman, holding the heights with Woronzov and Sacken and using 11,000 cavalry and Kleist's corps (under Winzingerode) to take Napoleon's right flank.

Before Sacken could get into position, Ney had established a presence on the plateau late on 6 March. Temporarily, at least, the initiative fell once again to Napoleon. His plans were well advanced and in the morning he proposed to launch Ney to the north of the enemy (supported by cavalry units from Grouchy and Victor), whilst Nansonty swung around to the south. Napoleon would hold the centre ground and pin down Sacken and Woronov by attacking the village of Heurtebise.

The engagement opened at 0900, with an artillery bombardment on Blücher's centre. Ney mistook this for the signal to launch his attack prematurely and it began to falter by 1100. Ney led more attacks from the front and gradually he gained ground. By now, Nansonty had struck and the enemy were beginning to fall back.

The battle swung back and forth and still Kleist had not appeared. Napoleon determined to destroy Blücher's centre and ordered the artillery and the Imperial Guard forward. Now Blücher began to withdraw, the pursuit ending at 2000; Kleist had got lost and never appeared. Sacken's 9,000 troops were never engaged throughout.

Blücher was furious; Winzingerode's attempted counterstroke had been an utter failure and with that he retreated towards Lâon. Napoleon had been fortunate, he had not been facing the rearguard at all, it was only luck which had saved him from potential disaster.

Arcis-sur-Aube, Battle of (1814)

Battle Name	**Dates**	**Campaign**
Arcis-sur-Aube	20-21 March 1814	Campaign in France
Troops Involved		
France	Day 1: 18,000 – 23,000 Day 2: 28,000 3,000 losses	Commanded by Napoleon
Allied	Day 1: 20,000 Day 2: 80,000 4,000 losses	Commanded by Schwarzenburg

Outcome:
Allied victory

The action at Arcis-sur-Aube was Napoleon's last throw of the dice. He believed that he could still dispute the Allied offensive against France and determined to strike at Schwarzenburg's army, which was covering a front of nearly 80 miles. At a stroke, Napoleon could cut Blücher off from Strasbourg and Schwarzenburg from Basle and it would also allow Napoleon to link up with the large French garrisons in Veroun and Metz.

Early on March 20th, Napoleon set off for Arcis-sur-Aube; his intelligence suggested that it was only held by a weak force of Bavarians, under Wrede, who formed the Allied rearguard. Schwarzenburg was not to be fooled by such a manoeuvre and was already marching to deal Napoleon a fatal blow.

Schwarzenburg knew that Sébastiani had crossed the Aube at Planey the day before (19 March). He assumed that Napoleon was heading for Troyes and, therefore, decided to concentrate his troops in the Troyes to Arcis area.

Napoleon ordered Marmont and Mortier to march first for Châlons and then for Epérnay, so confident was he that Blücher would not dare to continue his advance to Paris. Again, Napoleon would be mistaken. Confident of his plans, Napoleon sent Ney and Sébastiani to the south of the Aube to force Wrede to withdraw from Arcis-sur-Aube. He would continue on the north bank and rendezvous with them later for the march to Vitry and ultimately the Seine.

Remarkably, neither Ney nor Sébastiani ran into any opposition and they marched into Arcis-sur-Aube at 1100, even the bridge had not been completely destroyed. Napoleon joined them at 1300, having crossed the Aube at Torcy-le-Grand.

Suddenly, as it began to appear that an engagement would be a non-event, a large force of cavalry was seen coming from the area of Vanpoisson. Ney and Sébastiani could not halt them and it was only the timely arrival of Friant and his Old Guard division, holding the ridge that stemmed the tide.

At about the same time, Napoleon openly courted death by deliberately riding over an unexploded howitzer shell. His horse was killed, but the Emperor simply mounted another horse and started directing the troops.

The fighting became intense around Torcy-le-Petit and Torcy-le-Grand during the remaining hours of light and by nightfall it appeared that the French had the upper hand. Later that night, Lefebvre-Desnouettes arrived with 2,000 cavalry and Sébastiani managed to rout two enemy cavalry divisions, but he was forced back by Russian guns.

Little did Napoleon know that Schwarzenburg was massing some 80,000 men beyond the ridge to the south, stretching all the way from Voué (in the southwest) to Chaudrey (near the Aube in the northeast). Nonetheless, Napoleon felt uneasy and summoned MacDonald, who was advancing from Bray, to join him as soon as possible.

As far as Schwarzenburg was concerned, he could not be sure of Napoleon's strength and he was under pressure from the Tsar not to take unnecessary risks in his renewed offensive. In the end, Napoleon moved first, sending Sébastiani south to the plateau, followed by Ney. It was then that they saw the vast army Schwarzenburg had gathered. Now Napoleon knew he could not risk an attack without Gérard or MacDonald; the only option was to retire beyond the Aube. This manoeuvre would require more than the bridge at Arcis-sur-Aube and very hastily a pontoon bridge was created at Vilette. By 1330, the French were in full retreat. Oudinot and Sébastiani would cover the rear.

Finally, at 1500, Schwarzenburg shook off his lethargy and attacked. Oudinot and Sébastiani faced the oncoming, but not rushing, Schwarzenburg with great solidity. The fighting around Arcis-sur-Aube lasted until around 1800, when Oudinot's final rearguard fought their way to the bridge, destroying it as they reached the north bank of the river. Schwarzenburg did not attempt to pursue, he simply shadowed the French as they marched for Sompuis and joined up with MacDonald and Kellerman (who were near Ormes). Napoleon now switched his march to Sézanne and Frignicourt, still intent on reaching St Dizier and the garrison holding there waiting for him.

He arrived on 23 March, worse for wear but still able to summon up the fire to sweep aside some 8,000 cavalry that had ridden on an interception course from Vitry. Although he did not know it then, he had fought his last engagement of the campaign. On 28 March the Allies marched into Paris, Marmont and Mortier failed to regain the capital and at 0200 on 31 March they signed an armistice. On 1 April Napoleon had just 36,000 men at Fontainbleau; by 4 April he had 60,000. All of this was to no avail with 145,000 Allies in his capital. He had no option but to abdicate and finally signed his abdication document on 4 April 1814.

Ligny, Battle of (1815)

Battle Name	**Dates**	**Campaign**
Ligny	16 June 1815	Waterloo Campaign
Troops Involved		
France	80,000 11,500 losses	Commanded by Napoleon
Prussia	84,000 25,000 losses	Commanded by Blücher

Outcome:
French victory

Napoleon escaped from exile on Elba on 26 February 1815, by guile and good fortune he occupied Paris and began to reassemble his troops. The Allies reacted quickly, promising to restore Louis XVIII, but they were faced with coping with Napoleon with the troops already deployed. Inevitably, the initial responsibilities would fall on the 124,000 Prussians in Belgium and the 95,000 Anglo-Dutch.

The Austrians dealt with Murat at Tolentino on 2 May, whilst Napoleon planned to deal with the Allies one at a time, now with no other support. He marched his troops from Charleroi, hoping to drive a wedge between Blücher and Wellington. He would hold one of them with a smaller force and crush the other with his main army.

The natural choice of where the first blow should fall was predetermined for Napoleon as Blücher's Prussians were ahead of Wellington's army. He sent Ney with two corps to Quatre Bras to hold Wellington and sent Grouchy to attack Blücher at Ligny. Napoleon's reserve and the Imperial Guard would move to assist Grouchy and overwhelm the Prussians. It was then planned to force the Prussian survivors towards Namur before turning his attention to Wellington.

At 1430 on 16 June Napoleon launched his attack on the Prussian positions at Ligny. The enemy were holding a series of ridges and the town and whilst Napoleon struck at the front of the Prussian line, Vandamme hit their right. Napoleon had ordered d'Erlon to launch an enveloping attack, but the French commander was unsure whether this meant that he should be supporting Ney at Quatre Bras or Napoleon at Ligny.

Initially, this confusion gave Blücher a chance; gradually he was being forced back. Napoleon was about to launch the Imperial Guard at the Prussians when he saw new troops arriving on the battlefield. Again, there was confusion, Napoleon believing that they were Prussian reinforcements. The newcomers were, in fact, d'Erlon's men. By the time that the confusion was allayed, Napoleon had precious little time to launch his planned attack.

The net result of the delay still meant that the Prussians had been beaten, but they had not been annihilated as Napoleon had planned. In the dying moments of the battle Blücher was left for dead after taking part in a cavalry charge, but he and his men would return to haunt Napoleon. Ligny would be the Frenchman's last victory and an indecisive one at that.

Quatre Bras, Battle of (1815)

Battle Name	**Dates**	**Campaign**
Quatre Bras	16 June 1815	Waterloo Campaign
Troops Involved		
France	24,000 4,000 losses	Commanded by Ney
Britain / Holland /Germany	8,000 – 36,000 4,800 losses	Commanded by Wellington
Outcome:		
Allied victory		

Whilst Napoleon was dealing with Blücher's Prussians at Ligny, Ney's troops were despatched to capture the crossroads at Quatre Bras on the Charleroi-Brussels road. Wellington's Anglo-Dutch army was moving south from Brussels with Prince Bernard's Dutch in the lead. The Dutch had already reached Quatre Bras on the evening of 15 June when they encountered Ney's forward troops.

Ney did not strike that night and was curiously slow in developing his attacks the following morning, as it was he was faced with around 8,000 Dutch and could have easily swept them aside. Ney's troops finally deployed for a determined assault at 1400, led by Reille's corps of some 20,000 men. Despite strong Dutch resistance, he began to make headway.

By mid-afternoon, Wellington himself arrived on the battlefield, immediately sending his Brunswick corps in to support the Dutch. Ney had been relying on d'Erlon to help deliver the knock-out blow, but his corps was marching towards Ligny as a result of confusion in the orders. Ney still pressed the enemy, sending in Kellerman's cavalry to capture the crossroads. British infantry stiffened the centre of Wellington's line. They were under severe pressure from the French cavalry, some of the regiments not even having the opportunity to form up their squares to counter the attack. As the fighting intensified, time was running out for the French. Each

passing minute saw more Anglo-Dutch troops arrive, notably in the form of Picton and his command. These were still difficult times for Wellington, the Prince of Orange's troops were badly mauled by the French cavalry and on several occasions it looked as if Wellington's line would crack.

By the late afternoon, Wellington had some 36,000 men available to him, supported by 70 artillery pieces. At 1830 he was able to launch a full-scale counterattack on the faltering and exhausted French. The engagement was terminated at around 2100, by which time Ney had been driven from the crossroads and the Anglo-Dutch were firmly in control of the battlefield.

The battle could so easily have been won if Ney had been more proactive when he first encountered the Dutch at the crossroads. Equally, the opportunities still remained until at least midday. Even when d'Erlon was available to him, Ney made little immediate use of the reinforcements.

Wellington retired further up the Brussels road, covering his withdrawal as Napoleon's main army arrived. His chosen position was the village of Waterloo.

Wavre, Battle of (1815)

Battle Name	**Dates**	**Campaign**
Wavre	18-19 June 1815	Waterloo Campaign
Troops Involved		
France	33,000 2,500 losses	Commanded by Grouchy
Prussia	17,000 2,500 losses	Commanded by Thielmann

Outcome:
French victory

After the Battle of Ligny (17 June 1815), Napoleon had ordered Grouchy to pursue the retreating Prussians. Above all, Grouchy was to prevent a junction between the Prussians and the Anglo-Dutch army of Wellington and in addition he was to cover Napoleon's right flank.

Around 1200 on 17 June, Grouchy started off towards Gembloux, his progress impeded by poor weather and an imprecise knowledge of exactly where the Prussians were heading. At 1900, he reached Gembloux, where he decided to halt for the night. An hour later, he sent despatches to Napoleon, telling him that he believed that the Prussians had split into two columns. One, he understood, was heading for Wavre and the other to Liège. Grouchy considered the Wavre column to be the greatest threat and he intended to intercept it or force it away from Wellington.

Blücher had different plans and at 0400, he despatched Bülow to assist Wellington. Bülow's orders were explicit; he was to immediately attack Napoleon's right flank if he found him to be engaged with Wellington, otherwise he was to hold in cover in the St Lambert area until matters developed. Bülow would then be supported by Pirch who was to march next and subsequently by Zieten and Thielmann.

Grouchy got underway again on the morning of 18 June; later he would hear the battle at Waterloo in the distance, but determined to continue to Wavre and fulfil his orders. Grouchy received a despatch from Soult at 1530, reiterating his orders. Only Thielmann's troops still remained at Wavre and it was these that Grouchy engaged at 1400 on the eastern suburbs of Wavre.

Vandamme's corps had struck the Prussian pickets and outlying posts very hard, but he could not dislodge them from the two stone bridges over the River Dyle. Gérard arrived on the battlefield and Grouchy sent him toward Bierge to seize the bridge there, but his attack at 1700 failed. Grouchy now sent troops up-river to Limale and the small Prussian force here was overwhelmed by 1900; the French pressed on and captured Limelette. Around 1700, Grouchy had received orders from Napoleon to cover his flank, but as Grouchy was engaged he could do little.

Grouchy resumed the attack just after dawn and this time the Prussians were forced to retreat; he had his victory at 1000. Just half an hour later, he received word that Napoleon had been beaten at Waterloo. Thielmann already knew and it had seemed pointless to waste more lives.

Waterloo, Battle of (1815)

Battle Name	**Dates**	**Campaign**
Waterloo	18 June 1815	Waterloo Campaign
Troops Involved		
France	72,000 41,000 losses	Commanded by Napoleon
Britain / Holland / Germany / Prussia	68,000 – 140,000 22,000 losses	Commanded by Wellington/Blücher
Outcome:		
Allied victory		

Hard on the heels of Ligny and Quatre Bras, Napoleon had continued to press Wellington's army and that of Blücher's Prussians. At all costs, he needed to keep them apart and deal with them whilst they were isolated.

After Quatre Bras (16 June), Wellington had retired towards Brussels and on 17 June had set up his troops along a ridge south of Mont St Jean. Wellington's left flank rested on two farms (La Haye and Papelotte) and the village of Frischermont. The right was protected by Chateau Hougoumont and the centre by the farm of La Haye Sainte. The latter two defensive positions were ahead of his main line and he was determined to stand and fight. Wellington's two main corps (Picton and the Prince of Orange) were positioned behind the ridge, with his main cavalry force led by Uxbridge. Prudently, he had placed a strong force on his right to prevent Napoleon from outflanking him. Wellington proposed to fight a defensive action, taking advantage of the terrain and hoping to strike against isolated or over-exposed French units. Here he would wait for Blücher, secure in the knowledge that the Prussian would march to the sound of the guns.

Napoleon had detailed 33,000 men under Grouchy to keep the Prussians at bay whilst he dealt with Wellington. Blücher planned to form his army up around Wavre and support Wellington.

Napoleon's battle plan was none too subtle; whilst he held the village of Plancenoit on his right flank to protect against Prussian interference, he would launch a massive frontal assault on Wellington. His troops began to deploy during the early morning of 18 June with some 72,000 men and 246 artillery pieces, many of the latter deployed in a grand battery with which he intended to shatter Wellington's line. Wellington, cognizant of the fact that this was Napoleon's preferred tactic, deployed the bulk of his 68,000 men and 156 guns on the reverse of the slope.

The engagement got under way at around 1130, with Jérôme's division of Reille's corps attacking Hougoumont. This was intended as a feint as Napoleon hoped to draw off Wellington's reserves in an attempt to hold the position. Napoleon could not have accounted for the tenacity of the British Guards holding the chateau; they were to hold on to the location for the whole day. This brave defence had the reverse effect and gradually more and more French troops were thrown into attempts to take Hougoumont. It became something of an obsession and French attacks continued even after the strategic and tactical value of Hougoumont had ceased to be important.

At around 1330, Napoleon launched d'Erlon's corps at the ridge. He had already been informed that Blücher was coming up on his right and had formed up a defensive line near Plancenoit. Blücher would still take some time, he was still being harried by Grouchy, but only the Prussian rearguard at Wavre was actually being checked. The bulk of the Prussian army was heading for the battlefield, slogging along the muddy roads in the rain.

Although Grouchy was performing a useful function against the Prussians, he would be increasingly needed on the battlefield at Waterloo, but by the time he did turn and head for the main engagement, it would be too late.

Meanwhile, on the battlefield d'Erlon's columns were being cut to pieces by the British artillery, but they still came on and engaged Picton's infantry. Picton died in the midst of the battle, then Wellington ordered Uxbridge to launch a counterattack with his cavalry. Initially, at least, the charge was devastating; the French infantry were overwhelmed and were teeming back in disorder. The impetuous cavalry charge tore into the massed French batteries, then Napoleon's cavalry struck and by the time that the remnants of Uxbridge's command returned to the line, half of his men were gone.

At 1530 Wellington appeared to be withdrawing from the battlefield, which convinced Ney to launch an all-out charge with his cavalry. It was unsupported by either artillery or infantry and ultimately doomed. It appears that Ney took it upon himself to launch the attack. At this time, Napoleon was pre-occupied by the arrival of Blücher's troops at Plancenoit.

Ney's cavalry, numbering some 10,000 attacked on a front just over 700 yards wide. As the cavalry came on, the British infantry formed up in squares and shattered the cohesion of the French cavalry regiments with volley after volley. As the flower of the French martial elite was slaughtered, they had made an impact on Wellington's troops and in many places regiments were wavering. It was left to Napoleon to eventually order the withdrawal of Ney's cavalry at 1730.

With several Dutch and Hanoverian units deciding that enough was enough, Wellington was forced to commit the last of his reserves and hold the line. Like never before, the battle had been reduced to a war of attrition, but the pendulum was swinging Wellington's way, despite his perilous state.

On the French right, from around 1630, the Prussians had begun to arrive in ever-increasing numbers. Blücher, himself, had arrived and was personally directing the corps of Bülow and Pirch against Plancenoit. They took the village, but under the command of Napoleon the Young and Middle Guard retook the village at bayonet point. By now, Napoleon had committed his whole reserves, barring the Old Guard.

In the centre Ney was coordinating another attack; this time it would fall on the hard-pressed German Legion holding La Haye Sainte. French infantry gained a foothold into the farm and the

Prince of Orange saw two of his battalions sent up to support the farm slaughtered by French cavalry. The capture of the farm allowed Ney to bring up his artillery to closer range in order to rake the Allied line. As the French artillery opened up, the British regiments began to waver under the incessant bombardment. Only the positioning of Vandeleur's light cavalry prevented the infantry from being routed.

Had Napoleon had his reserves available at this moment, the outcome of the battle would have been very different. As it was, they were engaged against the Prussians and gradually cohesion and morale was restored in Wellington's exhausted and shattered regiments.

The only uncommitted troops were the Old Guard, but these were not thrown against the weakest position of Wellington's line. They headed west of La Haye Sainte and encountered the British Guards led by Wellington himself. The heads of the French columns were cut down barely 40 yards from the British positions and as more British units lapped around the French columns, the slaughter was immense. Another Old Guard attack had similarly failed against the Dutch. For the first time in their history the Old Guard had been defeated and with it Napoleon's hopes of victory had been lost.

Wellington now seized the initiative and ordered a general advance as Blücher's men overwhelmed Plancenoit. What remained of the Imperial Guard solidly formed a square to cover the retreat and was cut down. By nightfall, the age of Napoleon was over and the two victorious generals, Wellington and Blücher, met at Napoleon's former headquarters in the battlefield, La Belle Alliance. Napoleon would die, a broken and embittered man, in exile on St Helena on 5 May 1821.

Chapter 2

The Armies

Napoleonic soldiers, whether they were infantry, cavalry or artillerymen, are often portrayed as brightly uniformed automatons, marching in perfect formations on a battlefield, oblivious to casualties and heedless of danger. Discipline was, of course, important and the uniforms were certainly striking. Yet the armies that fought during the Napoleonic Wars were composed of men drilled in very different ways, with different outlooks on life and sometimes fortunate enough to be led by visionary characters that sought to ensure victory on the battlefield without having decimated their own men.

It would take a number of books to cover in any depth the various armies that fought in a wide variety of theatres during the Napoleonic Wars. This chapter aims to investigate the nature of five of the main protagonists: the Austrians, the British, the French, the Prussians and the Russians.

In the scenario books, which accompany this title, specific information can be found on some of the smaller nations and allies that fought in different theatres at various stages of the wars.

Austrian Army

At the time of the Napoleonic Wars the Austrian army was the third largest in the world. They had both the fortune and the misfortune to be at the very centre of Europe. It was fortunate in the sense that it could draw on troops from a wide variety of different cultures, but it had the misfortune to be embroiled, either directly or indirectly, in most of the conflicts. In fact throughout the period Austria was at war with France for 108 months. The French declared war on Austria in 1792 and the war would last for five years. They found themselves at war with France again in both 1799 and 1805, Austria being defeated on both occasions.

In the Napoleonic period Austria combined with Hungary had a population of 21,000,000 and in June 1799 they had over 90,000 men in Italy, a similar number in Germany and over 8,000 in Dalmatia. The Austrian army was certainly a huge force. For the Austerlitz Campaign in 1805 they had over 300,000 infantry, over 40,000 cavalry and over 12,000 artillerymen. By 1809 the Austrian army boasted up to 450,000 regular troops and upwards of 150,000 irregular troops.

Its major deficiencies were poor administration and staff work. It also had an enormous number of different nations fighting under the same banner and, above all, largely-useless generals. An Austrian infantryman was enlisted for six years and a cavalryman for ten. There was the potential for enormous confusion, with many of the non-German regiments unable to have any form of conversation or understand instructions.

The organization of the Austrian infantry can appear somewhat confusing. There were both German and Hungarian line regiments, there were grenadier and *jäger* (hunter) battalions as well as border infantry regiments known as Grenzers and also militia battalions. Each of the regiments could have between two and five battalions. The training of the infantry was somewhat inconsistent. Most of the regulars could be considered to be fairly well trained and disciplined, but many of the militia battalions lacked even the most basic training in musketry. Broadly, they were fairly well supplied and equipped and the supply system was relatively efficient.

Many Austrian regiments were more likely to be known by the name of their colonel, rather than their regimental number. Generally, the Austrian infantry was brave, if not particularly well led. On a smaller scale, Austrian regiments were perfectly capable of dealing with most tactical situations; it was in the larger battles where cohesion and mass movement were involved that they tended to fail.

The grenadiers were considered to be the elite and were often used as assault infantry. They were generally taller men and during most of the wars companies of grenadiers from each regiment were detached to form grenadier battalions. In many cases the grenadiers were held back as a reserve, to either reinforce or support a success, or to stabilize part of the line that was under pressure.

The Austrians also used jägers, or light infantry. They tended to be formed into battalions. In the later period the jägers forming the first and second ranks were armed with short muskets, whilst those in the third rank were given rifles. The jägers were some of the best trained units in the Austrian army and they performed exceptionally well in battles such as Leipzig. Jägers were adept at using battlefield terrain, where they would take up positions in cover to deliver volleys at close range.

The Grenzers were originally used to guard the border of the Ottoman Empire. They were initially raised to be militia and in wargaming terms they are an amalgam between line infantry and light infantry. They used skirmishing tactics and excelled in marksmanship. Grenzers were considered to be fairly low down on the scale of priorities for the Austrian army. They were often last in line to receive issues of weapons, uniform and additional training. However, at the Battle of Marengo in 1800 it was Grenzers that fought the rearguard actions. Some Grenzers eventually found themselves in French service between 1809 and 1814 after Austria was defeated in 1809.

The Austrians used a variety of different cavalry units. They were generally considered to be good horsemen that were well trained, with decent horses. The Austrians had cuirassiers, dragoons, *chevaulegeres* (light dragoons), hussars and *uhlans*. The French certainly considered Austrian hussars to be some of the best in Europe. Like the Austrian infantry, the cavalry was a broad church with many different nations. There were Hungarians, Moravians, Poles, Bohemians, English, Scots, Russians, Serbs, Prussians, French royalists and Croats. It was rare, however, to see Austrian cavalry operating as a mass force and they were certainly heavily outnumbered by Austrian infantry. Whereas the French usually had around three infantrymen to each cavalryman, it was closer to seven to one in the Austrian army.

A typical Austrian cavalry regiment had between two and four divisions, each with two squadrons. In 1809 Austria had 8 cuirassier regiments, 6 regiments of dragoons and the same number of chevaulegeres, 12 hussar regiments and 3 regiments of uhlans. They had a very similar number in the 1812 to 1813 period.

The dragoons and the hussars were supplied with carbines and other cavalry were often issued with rifles. The cuirassier's main weapon was a straight broadsword, but each also had a pair of pistols.

In comparison to the other major nations, Austria had twice as many cuirassiers regiments as Prussia, two to four less than Russia and four to six less than France. Austrian cuirassiers were distinctively clad in white coats and breeches. Austrian dragoons also wore white and could be distinguished from the chevaulegers who wore green jackets.

Austrian uhlans were armed with pistols, a curved sabre and a lance. There were also a number of men in each squadron armed with rifles or carbines. Theoretically, the uhlan was designed to deliver charges against unformed or scattered infantry units.

Austrian Hussars tended to be volunteers and the general rule was that the men were usually no older than thirty. They were considered to be some of the best light cavalry in Europe. Their distinctive uniforms were extensively copied by other European nations, including Britain, France, Russia, Spain and Poland.

In 1809 the Austrian artillery was organized into four separate artillery regiments, each with just under 3,000 men. Foot batteries tended to be either 6 or 12pdr, horse batteries 6 or 7pdr and there were also 7pdr and 10pdr howitzers. The lighter guns were invariably attached to infantry brigades, and the lighter horse artillery to the cavalry. The heavier guns were usually formed into artillery reserves.

The Austrians also employed engineers. Effectively they were sappers, trained to build and attack fortresses and erect field fortifications. In the period 1805 to 1809 the Austrians had nine pioneer divisions, a sapper battalion, a battalion responsible for erecting pontoon bridges and a miner battalion.

British Army

Although Britain had the largest and most effective navy in the world, it continually faced the prospect of seaborne invasion from the French. In fact during the Revolutionary period, the French tried to invade both Ireland and Wales. Britain, however, was a strong and wealthy country. It did not have a vast population, with just 18,500,000 compared to France's 30,000,000. Nonetheless, in terms of expenditure on its military machine, throughout the period Britain outspent France by three to one.

The British were consistently opposed to the French in every coalition from 1793 through to 1815. Britain often operated as the paymaster for the coalition facing France and was not always directly involved militarily.

The vast bulk of the British army were volunteers, whereas the officer corps tended to be aristocrats who had bought their commissions. In 1805 for example, the British army had nearly 162,000 regular troops, of which over 124,000 were infantry. They were, however, literally scattered across the globe. Fewer than 70,000 were in England, just over 30,000 in Ireland, 22,000 in the East Indies, 15,000 in the West Indies, 6,500 on Malta, 4,500 in Gibraltar and over 4,000 in Canada. Three years later when there were considerably more than 200,000 men under arms, 55,000 of them were in India.

The British army, however, was extremely efficient and well trained. Life for the soldier was based around his regiment. Even the French grudgingly admitted that the British soldier was both intrepid and skilful. They were often cool in battle and had excellent discipline. British infantry regiments, regardless of their origin or attachment to a particular town or county, were often packed not just with Englishmen but with Scots and Irish.

What can be particularly confusing about many of the British regiments is the fact that most of them only had a single battalion. This was not always the case. In 1809, for example, the 42nd Foot (later known as the Black Watch) had two battalions, whereas the 1st Foot (or Royal Scots) had four battalions and nearly 5,000 men. Each of the 104 infantry regiments were numbered in accordance with seniority. In effect each regiment was in fact an administrative formation and could consist of anything from one to as many as seven battalions. Having said this, British battalions were actually larger than either French or Prussian equivalents. At Waterloo in 1815, for example, the average size of a British battalion was 665 men, compared to a Prussian battalion of 610 and a French of just 520.

The British battalion had ten companies: eight centre companies, one grenadier and one light. The light and grenadier companies were supposedly the elite part of the battalion. Unlike many other nations, it was unlikely for the British to detach the grenadier companies to form battalions of grenadiers. It was not so much the case with the light companies, who were frequently detached to form a forward skirmish line.

Initially the British tended to have light and heavy dragoons, but after 1806 a number of the light dragoon regiments were converted into hussars. In fact, after our period, what remained of the light dragoon regiments became lancers.

The tactical unit for British cavalry was the squadron and at Waterloo the squadrons ranged from around 112 to 180 men; larger than their French equivalents. A cavalry regiment consisted of around five squadrons, but one of the squadrons was always left back as a depot at their home station. This meant that most of the British cavalry regiments could actually muster between two and four squadrons in the field. On average, a cavalry regiment would have around 460 men. There were exceptions to this and some regiments were considerably larger. There were two or three regiments in a brigade. This meant that at Waterloo, for example, some British regiments were actually bigger than a French brigade.

At Waterloo, Wellington had four guard cavalry regiments that mustered nine squadrons, three dragoon regiments also with nine squadrons, nine light dragoon regiments with thirty squadrons and ten hussar regiments with thirty-five squadrons.

Not all of the cavalry regiments used at Waterloo were strictly British: four were from the King's German Legion, one was Hanoverian and eight of the other regiments were from British allies.

The elite of the British cavalry were the 1st and 2nd Life Guards and the Royal Horse Guards. The most numerous, however, were the two different types of dragoon.

Unlike the French, even the Dragoon Guards did not wear a cuirass. The light cavalry were a mix of both light dragoons and hussars. Originally the dragoons were mounted infantry and had been used for patrols and reconnaissance. But they were soon seen as having a distinct value as being effective in close combat.

Another significant portion of the British cavalry was those known as the King's German Legion. France had occupied the electorate of Hanover in July 1803, but the ruler of Hanover was England's King George III. Although Hanoverians were forbidden to take up arms against the French, thousands flocked to England to fight for their monarch. By 1805 they provided a number of light and heavy cavalry regiments, artillery batteries, two light infantry regiments and eight line infantry regiments. They were supplied entirely by the British.

British artillery of the period was extremely well trained and efficient. It could trace its roots back almost a hundred years. It was, however, the smallest of the three parts of the army. At Waterloo, for example, the British had just five foot and eight horse batteries. Typically a battery would have six artillery pieces and the battery would be divided into three divisions, each with two guns. Unlike some other armies, notably the French, the British tended to space their artillery out along the battle line. Artillery was subordinate in the British army and intended to support the infantry. By contrast, the French, for example, would mass their artillery and concentrate their fire on a particular portion of the enemy's frontline in order to achieve disruption and create the opportunity for a breakthrough.

The most commonly used British artillery piece of the period was the 6pdr and certainly it was the most numerous, although British artillery could be 9, 12 or 24pdrs. The problem with the heavier guns is that they were less manoeuvrable on the battlefield and needed more horses. The British also used rockets. In fact they were a modification of a weapon that had been used against them in India by the Mughals. Congreve had modified the weapon by 1808 and had created six different calibres, all of which had a steel case, a conical or spherical head and were filled with gunpowder. They were fired from a trough and could actually outrange the bulk of artillery pieces of the time. Although not all British commanders were convinced that they were particularly useful weapons, they were in fact deployed at Waterloo and elsewhere.

French Army

The French reached the height of their power during the reign of Napoleon. Their military machine was certainly one of the most powerful and effective, yet ultimately it was defeated. Many countries tried to emulate the French and use their military expertise, yet the army can trace its successes back to the periods before the rise of Napoleon, when successive French kings mobilized huge armies of well-paid, well-disciplined troops that fought wars across the world.

Before Napoleon, the French army was a royal one and the French maintained the largest standing army in Europe. In 1762, for example, they had 187 line infantry battalions, each with over 600 men. It was this force that formed the backbone of the army that fought in the Seven Years War, which turned out to be a disastrous conflict for the French. The war had also seen them lose much of their influence in North America. It was a war that they fought at a great disadvantage; as primarily a land power the French found it almost impossible to reinforce their outnumbered colonists, particularly in Canada.

By means of compensation for their loss of Canada the French were instrumental in assisting the rebels in the American colonies to gain their independence from the British. The final major engagement at Yorktown saw a combined American and French force overcome the British.

The French army was not immune to the enormous repercussions of the French Revolution that began in 1789. Revolutionary France required every able-bodied Frenchman to be available for military service. France was the first country to use universal conscription. To begin with it was a poor force. The regular troops were demoralized and the rest of the army consisted of

poorly-trained volunteers. In 1790 there had been a series of mutinies and insubordination was rife. In 1791 the French infantry was reorganized into 104 line regiments and 12 light infantry battalions. But this was still under strength and battalions of National Guardsmen were called up as reservists.

One of the other major problems was that many of the officers were royalists and those that did not meet their end at the guillotine fled the country. Nonetheless, changes were still underway and successive groups of National Guardsmen were called up in the early 1790s. The trick was to turn these recruits into effective units.

By the early 1800s around 100,000 conscript troops were called up each year. The size of the French army was reaching its peak between 1803 and 1807. Napoleon was making root and branch changes. Over 170 senior officers were retired due to their age or incompetence and the French army numbered at least 450,000.

Even Napoleon's great victories, such as Ulm, Austerlitz and Jena cost the French army dear in killed and wounded. At the Battle of Eylau upwards of 25,000 French became casualties. All of these losses had to be made good with equally well-trained men. However, when the new war with Austria broke out in 1808 the French army was in very good shape and there was still an emphasis on drill and manoeuvre.

By 1808, around 200,000 Frenchmen were called up for the annual conscriptions. Many of these conscripts, who were inferior soldiers in many ways, found themselves fighting in the Peninsular War and things became even worse during this time when Napoleon started to withdraw many of his best troops for his impending invasion of Russia.

Napoleon had time to retrain his forces. The soldiers that marched on Russia were well trained, clothed and armed. The cavalry numbers were swelled by Polish and German troops. The Grand Army that Napoleon assembled for the invasion of Russia was the largest he had ever commanded. It would require much greater skill than ever before, not just on the battlefield, but to ensure that they were well supplied. This was not always easily achieved and by the time the French reached Moscow half of the army was gone. Borodino cost the French very dear with 30,000 killed or wounded, but it was not the battles that necessarily decimated the French army, it was the weather, the lack of supplies and the constant harassment by irregular Russian units, such as Cossacks. By the time Napoleon ignominiously retreated from Russia, barely 90,000 men remained of the 680,000 that had begun. At least 370,000 were dead and around 200,000 were prisoners.

By 1813 Napoleon, in order to wage a war in Germany, needed virtually an entirely new army. He amassed a huge force but many of the soldiers were young and inexperienced. Very few of the officers could have been considered anything more than capable.

For the campaigns in France in the following year, 300,000 men were conscripted and additional call-ups brought the National Guard up to 112,000. This entire force had to deal with an enemy of upwards of half a million men and at no time during this campaign did Napoleon muster a field force of more than about 75,000. The army by this stage had been worn down. Paris was taken and Napoleon was forced to abdicate.

It was remarkable, however, that when he returned for what has become known as the 100 Days Campaign, he managed to amass a force capable of facing the allied occupation forces in Western Europe. Many of the men had had at least one campaign behind them, but by and large the French discipline had gone. Men lacked uniforms, shoes and equipment, the French cavalry was a shadow of its former self and many of the senior officers were of dubious loyalty.

The strength of the French infantry varied over the period. When Napoleon came to power, France had 90 line and 26 light regiments. By 1813 this had been built up to 137 line and 35 light. It dropped back again for the Waterloo campaign to 90 line and 15 light regiments. Theoretically, a regiment had between two and five battalions. In the period 1808 to 1815 battalions would have six companies: four fusilier companies, one of grenadiers and one of *voltigeurs* (skirmishers).

The line infantry was trained to carry out a wide range of different manoeuvres, which included skirmishing. However, the French deployed specialist light infantry for scouting and for movement in advance of the main force. They were considered to be particularly effective and were viewed as being superior to many other light forces used by other nations during the period. Light troops would not be adverse to fixing bayonets and storming positions, regardless of whether they were supported by line infantry or not. The French formed units of irregulars, in the Peninsula, for example. They were largely recruited to deal with other irregular troops, smugglers, partisans or to fight in particularly difficult terrain, such as frontier regions or mountainous countryside.

Although each infantry battalion had a grenadier and a voltigeur company, the bulk of the battalion was made up of fusiliers. They were known as *chasseurs* in light infantry battalions. Effectively the fusiliers or chasseurs were the centre companies of the battalion. It was Napoleon who introduced the concept of the voltigeur. They were trained to fire more rapidly and accurately than a fusilier and were also often detailed to carry out skirmishing and scouting duties. Voltigeur companies could often be detached from the battalion to form a larger formation of their own.

On the other hand, the grenadiers, or *carabiniers* in light infantry battalions, were considered to be the elite troops of the battalion. They were veteran men, usually with more military service than any other soldier in the battalion. They would often be used as the spearhead of the attacking force. Effectively they were crack troops that were paid more money and were often used for close assault work.

There were, of course an enormous number of foreign troops that fought in the French armies. This is in addition to foreign soldiers who enlisted directly into French regiments. Typical examples of exclusively-foreign units include the Swiss and Croatian regiments, the Polish Vistula Legion or the Portuguese Legion. There were also troops from allied or dependent states, such as those from Westphalia, Bavaria or Saxony. In 1803 northern Italians were recruited and they would later become the 11th French Regiment, in 1811. Corsicans also served in the French army, the original battalion being raised in 1803. The Swiss had been used primarily as mercenaries in the period before Napoleon. In fact it was Swiss Guards that defended key positions in Paris during the French Revolution. The Polish Vistula Legion served in most of the major campaigns and four regiments were even attached to Napoleon's guard. The Croats provided at least four provisional regiments, the first one being formed in 1811, and they were to serve in Russia, but all of the regiments were disbanded in 1813.

The Irish Legion was a particularly interesting development. A legion had been created in 1803 in the hope that it would be instrumental in assisting in the invasion of Ireland. By 1809 the legion had five battalions; the first two being almost exclusively Irish and the other battalions were a mixture of different nationalities. The Irish Legion became the 3rd Regiment Etranger in 1811 and by that stage it had served in the Peninsular War.

There were a wide variety of other flamboyant regiments made up of foreign nationals. Some were composed of deserters from other armies, such as the Prussians, whilst others consisted of Germans, British, Swedes, Greeks, Hungarians, Danes and Spaniards. From time to time, as the campaigns ebbed and flowed across the continent, other units were created, such as the Albanians that served between 1808 and 1812.

The structure of the French cavalry was somewhat more complicated than any of the other nations of the period, as there was a wide variety of different types of unit. The paper strength of a French cavalry regiment ranged between 800 and 1,200 men, but in reality it could be as little as half of this. Cavalry was far more difficult to replace both in terms of equipment and in training. Most of the regiments would have three or four squadrons and each squadron would consist of two companies. Initially the quality of the French cavalry had dipped dramatically, as it was the favoured military arm of the French nobility and they had been affected and purged by the revolution.

Prior to the campaigns in 1805 and 1812, French cavalry was intensively trained and brought up to battle readiness and kitted out with the best uniforms and the best available mounts. In extended wars of attrition, such as the Russian campaign in 1812, the cavalry suffered even more extensive losses than the infantry. At least 175,000 trained cavalry mounts were lost during the campaign. These were extremely difficult to replace with any kind of quality mount. Northern

France was the area where most of the better French horses were bred, although huge numbers came from Hungary, Poland and southern Russia.

In theory, each squadron would have a specific coloured horse. For example the 1st Squadron would tend to be given black horses, whereas the 3rd Squadron would have chestnut horses.

Horse Carabiniers were originally raised in the late 17th century, however during the Napoleonic Wars there were only two regiments. The idea was that they were mounted men that should have the capacity to fight on foot when needed. In 1803 the two regiments only had two squadrons each, but Napoleon strengthened the regiments and brought them up to three and four squadrons respectively. In 1809 the units were given armour. In 1805 they were given dragoon muskets and by 1812 they had been given shorter cavalry carbines. They suffered badly during the Russian campaign and in fact by 1815 some of them had deserted to the British and there were enough men to form a troop known as the Bourbon Cavalry Corps.

The cuirassiers were direct descendents of the medieval knight. They were invariably huge men on massive horses, wearing armour and considered to be elite troops. Napoleon created fifteen cuirassier regiments during his period in power. They used their weight and brute force to turn the fortunes of a battle. The Great Redoubt at Borodino was captured by French and Saxon cuirassiers; such was the power of this type of cavalry. They wore long, black boots to protect their legs, had steel helmets with a brass cone and a steel cuirass with front and back plates. Theoretically they would have pistols as well as long, straight sabres, but in reality very few had pistols and those that did carried little ammunition.

French Dragoons were trained to operate as both infantry and cavalry. The original idea was that a large number of infantry could be moved rapidly on horseback during a campaign. In the period 1799 to 1800 the French had twenty dragoon regiments. By 1804 there were thirty. It is significant to note that by 1811 the trend was on the decline, as six regiments had been converted into lancer regiments and by 1815 there were just fifteen dragoon regiments left. By 1807 dragoons had been relegated to secondary theatres of war, such as Italy and Spain. They were a great success in Spain, able to move considerable distances and then have the ability to deal with situations due to their fire power from carbines and their willingness to engage in close combat. In 1805, due to the difficulty in finding sufficient horses, 6,000 Dragoons were organized into four foot dragoon regiments and they would be assigned the duty of guarding the artillery and the baggage trains. Towards the end of the period dragoons would most commonly be armed with straight sabres and with muskets, rather than light cavalry carbines. This gave them a better range.

The first major type of truly light horse was the lancer, modelled very much on the Polish lancer. Before the Russian campaign Napoleon had converted a number of dragoon regiments into lancer units. Additional lancer regiments were created from Polish troops. They were to be of particular value from 1813 to 1815, where they fought with effectiveness at Dresden, Leipzig and Quatre Bras.

Another type of French light cavalry were the horse chasseurs. In 1798 there were twenty-two regiments, twenty-four in 1804, thirty-one in 1811 and just fifteen in 1815. It is particularly significant that a large number of foreign troops served in this branch of the cavalry, including Belgians, Italians and Germans. Chasseurs were a strange hybrid: they could operate on foot yet perform many of the functions of Hussars. They were used for reconnaissance work and for campaigns where large-scale battles were rare. They were armed with a cavalry sabre; originally two different types, one specifically designed for themselves and the other basically a Hussar sabre. Later these were replaced with a slightly-curved blade of specific design. They were also given carbines and bayonets.

Hussars had become popular as a mounted force towards the end of the 17th century. A hundred years later, in 1798, the French had twelve hussar regiments. Their total strength would never grow far beyond this and by 1815 there were only seven left. Hussars considered themselves to be better swordsmen and horsemen than any other French cavalry. They were used as scouts and as troops to screen a main body but they were also used in pitched battle.

Unlike many other nations' artillery, the French gunners were often as aggressive as other wings of the army and also more prepared to be deployed forward of the main battle line. In 1805 there were eight regiments of foot artillery and six of horse artillery. Napoleon, with his background in artillery, poured considerable time, money and effort into improving French artillery and the number of guns used in battle increased every year. There were just 2 guns to every 1,000 men at Austerlitz; it reached a pinnacle of 4.5 guns to each 1,000 men at Borodino, falling back to 3.5 at Waterloo. Napoleon wanted literally thousands of cannon to be made available to the army. In 1809 he calculated that the army needed 6,000 artillery pieces.

It was not necessarily the quality of the French gunners that made the difference on the battlefield. Napoleon favoured using artillery offensively, rather than the approach taken by most of his opponents, which was effectively to defend their cavalry and infantry. The French would make great use of flanking fire and the French gunner was extremely bold and

would move forward to ensure that each shot caused maximum damage and disruption. In some cases the artillery would be brought to within as little as 60 yards of the enemy front line. The French also used artillery as a massed battery (or grand battery), firing at a decisive point in the enemy line. Sometimes opponents would try to counter this by massing their own artillery and engaging in a counter-battery fire duel.

After the invasion of Russia in 1812 the quality of French artillery dropped considerably, with the loss of at least 1,200 cannon. By the Waterloo campaign it was not so much the lack of cannon, but the drop in the number of trained gunners and good horses that was telling. French gunners needed to be well trained; they had a multiplicity of roles. Not only did they have to serve the guns, but they also had to be part of an important supply train, deal with horses and have responsibility for ongoing repairs.

Like all artillery, the caissons and wagons were pulled by horses. A 12pdr foot cannon would need twelve horses just for the gun and it would need three caissons, each with two more horses. If possible the French would double the amount of horses made available to speed up movement.

Like most armies, there were effectively two branches of artillery; the foot and the horse. A foot artillery company or battery would consist of up to 120 men with six cannon and two 6" mortars. Horse artillery, needing greater mobility and more horses, were relatively more expensive. By 1810 there were seven horse artillery regiments, each with three squadrons and each squadron having two companies.

A vital component of the artillery was the artillery train. They would ensure that sufficient ammunition was available for the artillery. A 4pdr caisson, for example, could carry twenty-six cannonballs, with twenty-six large and twenty-four small cartridges. Alternatively it could carry 12,000 infantry cartridges. An 8pdr would normally have two ammunition wagons with just over 100 projectiles available to it. But Napoleon often insisted that the artillery train provided each artillery piece with at least double that amount.

No mention of the French army can be complete without referring to the guard troops, of which there were infantry, cavalry and artillery. In theory they were all considered to be elite troops, Napoleon's personal army. Initially, very few of them were actively used on campaigns, but by 1813 things had changed and they became Napoleon's shock troops. The guard troops were better paid, they had better housing, and they had the best uniforms and equipment, but were subjected to the strictest discipline.

In 1804 Napoleon set about converting the existing Consular Guard into the Imperial Guard. The men that would form the Imperial Guard would be drawn from every regiment of infantry, cavalry and artillery. There were strict rules governing eligibility to join the Imperial Guard. Men would have to have been in service for at least five years, exceed a minimum height requirement and have served in at least two campaigns.

As far as the infantry was concerned, three distinctive types of guard units would develop. The Young Guard appeared in 1809. The Middle Guard was a smaller force and only a few regiments and battalions were ranked as Middle Guard troops. The most famous, of course, was the Old

Guard. In reality there were only really two regiments of the proper Old Guard: the 1st Regiment of Foot Grenadiers and the 1st Regiment of Foot Chasseurs. They were considered to be the cream of the army, battle hardened, physically strong and bold. They fought in innumerable battles.

The guard cavalry was similarly complicated and consisted of horse grenadiers, elite gendarmes, dragoons, chasseurs-a-cheval, Mamelukes, lancers and scouts and honour guards. Effectively regiments of each of these types were generally assigned to each of the three types of guard, with the Old Guard receiving the larger and more senior regiments and the Middle Guard tending to be the last in line to receive cavalry units. A similar system was used to recruit men into the guard cavalry units. For example, in 1813 Napoleon requested that each cavalry regiment currently operating in Spain send their best twenty men to become part of the Old Guard.

The Horse Grenadiers were the most senior regiment. They had been former Consular Guards and were brought into the Imperial Guard in 1804. The Horse Grenadiers attached to the Old Guard invariably rode large, black horses. The Horse Grenadiers attached to the Young Guard rode brown horses.

In 1801 the elite Gendarmes were created. Their primary duty was to handle security around the army headquarters and to guard and escort correspondence, VIPs and prisoners. They too rode large, black horses and had slightly smaller bearskins than the Horse Grenadiers.

The Regiment of Guard Dragoons was created in 1806 and they would fight in a number of engagements, including Borodino and Leipzig. They were armed with slightly-curved sabres and dragoon muskets and they tended to ride chestnut horses.

The first Regiment of the Horse Chasseurs was created to become Napoleon's own personal escort. They were created for very practical reasons, as in 1796 Napoleon had almost been captured by Austrian light cavalry at Borghetto and only escaped by climbing over a wall. Chasseurs wore essentially Hussar-style uniforms and they were armed with curved sabres, pistols and carbines.

An equally colourful unit was the Guard Mamelukes. They were created in 1802. Many of the men were actually Turkish, Georgian, Egyptian or Greek and carried a pair of pistols, a sabre, a dagger, a mace and later a battleaxe. By 1815 the vast majority of the men were in fact French, although they still wore their characteristic turban.

In 1807 Napoleon raised a guard regiment of Polish light horse; effectively they would be lancers. They were expert horsemen and adept at maintaining their formation on the battlefield. By Waterloo there was only a single squadron of the Polish Lancers left. The second regiment of lancers were in fact Dutch and were formed in 1810. They had one of the most striking uniforms; largely red, earning them the name Red Lancers. There were five squadrons in the Young Guard, but these wore blue jackets faced with red. In 1812 the regiment numbered slightly less than 700 men and after the Russian campaign the regiment was rebuilt and maintained its strength until at least 1814.

In 1813 Napoleon created three new regiments for the guard cavalry; effectively they were to be French versions of the Russian Cossacks. These men were armed with lances and sabres, whilst others had carbines and sabres. The three regiments did not last very long as they were disbanded in June 1814.

The final component of the French Guard cavalry was the Honour Guard. These were men in their twenties drawn from wealthy families. In 1813 there were 15,000 of them, organized broadly into four regiments. Originally they were named Life Guards, but many of them turned out to be less than enthusiastic about war and deserted and instead they were called the Honour Guard.

The final component of the Imperial Guard was the artillery. In 1800 there were just twelve artillery pieces but over the next few years several foot and horse batteries would be created. The horse artillery was considered to be elite; they had the best horses and best gunners in France. By 1813 the Old Guard had forty artillery pieces and the Young Guard thirty-two. It too was supported by its own artillery train, effectively ammunition wagons. Theoretically, on campaign one battery of artillery was teamed up with one company from the artillery train. This was designed to ensure that each battery had sufficient draft horses, limbers, ammunition and stores to maintain its effectiveness as a unit.

Prussian Army

In 1740 the Prussians had the fourth largest army in Europe, with 85,000 men. It was an admirable military machine that had been fashioned by Frederick the Great. In the 1750s he had embarked on a series of conquests but ultimately it had led to Prussia having to face the combined strength of France, Austria and Russia. Prussia's allies were Britain and a handful of German states. The Seven Years War ended in victory for Prussia, allowing them to enlarge their territories and become an equal to the other European world powers.

Unfortunately, after Frederick the Great's death the army began to slump into something of a decline and this would culminate in their defeat at the hands of the French in 1806. In 1806 the Prussian army could muster 200,000 men, of which 133,000 were infantry. The army would be defeated, however, at Jena and Auerstadt. Prussia had gone into the war with enormous confidence, but this had been rudely shattered on 14 October. Tradition and past glory was no substitute for military prowess and the weapons that the Prussians were using were probably the worst in Europe.

At Jena the Prussians lost a staggering 25,000 killed or wounded. In the ashes of humiliation the Prussian army was reformed in the period 1807 to 1812. The peace treaty of 1808 restricted the army to little over 40,000 men yet the Prussians realized that serious reform needed to be

undertaken in order to modernize the army. By 1812 the army was still small but certainly far more effective. The Prussians had come up with a clever system. They called up a number of recruits, which they assigned to regiments. The men were given basic military training and then discharged, so that another batch of recruits could be called up. This assured that they would never go above the ceiling imposed upon them in the treaty. In effect this meant that the Prussians had developed men capable of creating reserve battalions.

By 1813 the Prussians had regular troops, reservists, volunteers and also the Landwehr. These were effectively a national militia. Regiments were formed into brigades, each of which had their own infantry, cavalry and artillery. As part of the secret system of building up the army, brigades were reinforced with additional units but still called brigades even though they were in fact divisions. The creation of Landwehr units also accelerated.

In 1813 each of the six districts of Prussia raised between 11 and 68 battalions of infantry, the equivalent of 3 to 17 regiments, and between 8 and 40 squadrons of cavalry, the equivalent of 2 to 10 regiments. Simultaneously, volunteers were called up, many of them using hunting rifles as their primary weapon. Most were infantry, whilst others formed cavalry units.

The Prussians also had Free Corps units. They consisted of men drawn from across Germany. In 1813 this force amounted to 2,900 infantry, 600 cavalry and 120 artillerymen. The corps was later dissolved, in 1814, with what remained of the men being integrated into regular units.

The period saw an enormous expansion of the Prussian army. By August 1813 there were 90 battalions of regular infantry, 39 battalions of reservists and garrison infantry, 8 battalions of jägers and foreign troops and 151 battalions of Landwehr.

The cavalry was similarly impressive, with 89 squadrons of regulars and volunteers, 22 squadrons of reserve cavalry, 23 jäger and foreign squadrons and 113 Landwehr squadrons. The artillery boasted 50 field batteries and 33 fortress and siege companies. By 1815 some 279 infantry battalions, 280 cavalry squadrons and 78 artillery batteries were organized into 6 army corps and a Guard corps.

The Prussian infantry that faced the French in 1806 was remarkably similar to that which had fought for Frederick the Great, having barely changed since the Seven Years War. There had to be a rapid series of changes if the Prussian army was ever to match its opponents on the battlefield.

The Prussian infantry retained its characteristic three-line system. In an 1812 to 1813 Prussian brigade of two regiments, the front line were fusiliers, effectively light infantry, of which there would be two battalions. Behind them were three musketeer battalions, in essence line infantry, while the third line consisted of another musketeer battalion and a combined grenadier battalion. There were a few slight changes by 1815. The Prussian brigade now tended to comprise nine battalions from three regiments. It retained the two fusilier battalions as the front line, but now had four musketeer battalions in the next and finally two more musketeer battalions and a fusilier battalion forming the third line. The grenadiers had been effectively stripped out to form their own regiments and brigades.

The Prussians also had their own foot guard regiment, with one fusilier and two musketeer battalions. By 1813 the 1st Foot Guard Regiment retained its three battalions and there was also a 2nd Foot Guard Regiment, a 450-man Guard Jäger Battalion and a similar number of men in a Guard Schutzen Battalion.

As already mentioned, the grenadiers were eventually stripped out of the line regiments to create their own regiments and brigades. Back in 1806 the Prussians had twenty-seven grenadier battalions. By the end of 1814 grenadier battalions were formed into full regiments for the first time. Unlike many other grenadier units in other armies, they were not held back but often attached to the advance guard. This meant that the grenadiers often saw a great deal of fighting.

The Prussians had a wide variety of different light infantry, including the fusiliers, jägers and schutzen. The jägers were armed with rifles and considered to be something of an elite within the infantry. The fusiliers, unlike the jägers, were armed with an ordinary musket and the schutzen were designed to fight rather like jägers, but were mainly armed with smoothbore muskets instead of rifles. The Landwehr were originally designed for home defence and were initially armed with discarded French muskets that had been captured by the Russians. They varied in terms of quality. Some of the Landwehr were as willing as regulars to fight on until they ran out of ammunition, whilst others were rather less reliable. Their uniforms were simple and functional and tended to be either a black or dark blue coat with grey or white trousers.

The Prussian cavalry often distinguished itself with bold charges and a willingness to prevent disaster by covering the retreat of the Prussian army. In 1805 the Prussians had thirteen cuirassier regiments, fourteen dragoon regiments and nine Hussar regiments. This was significantly reduced after the defeat in 1806. However the force was not significantly larger even by 1815, when there were just five cuirassier, four dragoon, thirteen hussar and nine uhlan regiments. In theory, each of the cavalry regiments had four squadrons, but many only had two or three. There were also guard cavalry. In 1813 for example, there was the Garde du Corps and the guard light cavalry. Both regiments had four squadrons. Three new regiments were created in 1815: the Guard Uhlans, Guard Dragoons and Guard Hussars.

The most prestigious Prussian cavalry were the cuirassiers and the regiments got the pick of men and horses. They were perfectly capable and willing to charge en masse straight into the enemy on the battlefield.

Prussian dragoons, although of lower status, tended to find themselves in the thick of battle more often. Dragoon regiments often had horse volunteer jägers attached to them. The dragoons were armed with a curved sabre and a number of men in each squadron were given cavalry carbines. The uhlans were lance-armed light cavalry and they were also given a curved sabre. Many of the men were also armed with pistols.

Prussian hussars also saw a great deal of action. They were hard-riding and hard-fighting men, armed with curved sabres and carbines and they wore perhaps the most expensive uniforms in the army. Prussian hussars were never especially numerous, on average certainly numbering no more than a dozen regiments. Landwehr cavalry tended to be light and largely either hussars or uhlans, though their lack of training made them particularly-poor lancers.

In the period up to 1806, Prussian artillery had been fairly neglected as an arm of the military. After the 1806 defeat the Prussians were restricted to an artillery force, including engineers, of just 6,000 men. After 1812, enormous amounts of time, energy and money were put into building up the army's artillery fire power. Between March and August 1813 the number of field pieces available doubled to around 400.

These 400 field pieces were organized into fifty batteries. Thirty-eight were foot batteries and the remainder horse batteries. There were other, smaller numbers of guns available, such as to the Guard horse and foot batteries, the Berlin Landsturm and the Free Corps. An additional five howitzer batteries were created in time for the Waterloo campaign. The artillery would tend to be used to engage the enemy artillery and to cover the rest of the army while it deployed.

Russian Army

By European standards Russia was a relatively-young country and had only come to any kind of prominence at the beginning of the 18th century. Russia had played an important role during the Seven Years War and, fortunately for Prussia, had switched sides with a change of monarch after the Russians had been fighting the Prussians for almost a decade.

After the Seven Years War there was a massive purge throughout the army. An enormous number of senior officers were fired due to widespread fraud in the armed forces. During the Napoleonic Wars the monarch was Tsar Alexander I. He had come to the throne after the murder of his father. Russia's growing prominence could not keep the country out of the Napoleonic Wars for very long. In the early years the Russians were campaigning in Persia and against the Turks. But as France began to expand and their influence was felt more keenly, Russia allied itself with Britain and Austria against Napoleon. In 1805 Russia went as far as signing an offensive alliance against France.

The Russian army was in many respects true to the stereotypical image of the poorly treated serf, commanded by aristocratic officers. After the Austrians had been defeated in 1806, France agreed to help Russia against Turkey, whilst Russia pledged to become engaged in the war against Britain. The brief alliance would bring sweeping changes to the Russian army and prepare it for the enormous struggle it would soon find itself thrown into when Napoleon invaded Russia in 1812. In fact, when the French invaded their force was twice as large as the existing Russian army.

Although we largely remember the French victories on Russian soil, there were a number of major Russian victories, which helped to ensure that Napoleon's invasion was ultimately unsuccessful. As 1813 dawned, Russia opened a campaign against France on its own, although it was later joined by the Prussians and the Austrians. Many credit the Russian victories in 1812 for instilling belief into the Austrians that they could beat the French despite their repeated failures to do so in the past.

The Russians had to cope with enormous supply and administrative problems. It was 1,000 miles from the battle fronts in Germany to central Russia and everything had to be moved by wagon. Russian infantrymen were relatively self-sufficient. They carried some food with them but also relied heavily on foraging.

The Russian army remained comparatively small throughout the Napoleonic period. Russian infantry regiments tended to have two or three battalions, cavalry regiments between four and ten squadrons. In 1804 there were three to four guard infantry regiments, thirteen grenadier regiments, eighty-four musketeer regiments and twenty-two jäger regiments. In 1805 the cavalry could muster four regiments of guard cavalry and forty Cossack regiments.

During the war of 1812 there were three western armies, mustering up to 240,000 men. The Army of the Danube had some 60,000. There was a corps at Riga of 40,000, one half its size in Finland and two reserve corps with a total strength of no more than 70,000. Discipline was harsh, although officers were encouraged to behave like fathers to children. Punishment and torture was abolished, but beatings were common.

The vast majority of Russian infantrymen were illiterate peasants. They were well supplied with alcohol, which would aid their willingness to fight. The Russian army also put great store in the use of the bayonet, believing it to be a true Russian weapon and far more decisive than the musket. However, inferior-quality weapons plagued the Russian army. Various calibres of musket were used and the Russians purchased muskets from a wide variety of sources.

In 1800, Russia had a western population of some 35,000,000. The army around these times, as far as infantry was concerned, comprised of grenadiers, which were heavy infantry, musketeers, which were the line infantry and jäger regiments operating as light infantry. Many of the regiments began the 1812 campaign at more or less full strength, but by 1813 the Russian infantry had suffered enormous losses. Many of the regiments had a handful of veterans, but most of the men were virtually raw recruits rushed in to bring the regiments up to strength. In theory the infantry regiments should have had two or three battalions, but invariably they only had one. Grenadier regiments received priority and on average they could muster two battalions and guard units could have as many as three.

In order of seniority and performance, obviously the best infantry were the guards, then the grenadiers, the jägers and finally the musketeers. There were some very specific and recognizable units within the infantry of the Russian army, such as the famous Pablovsk Grenadier Regiment, who in 1812 and beyond still wore the mitre cap.

In times of dire need, such as the low points in the 1812 campaign when the French managed to penetrate as far as Moscow itself, militia units were created and armed with pikes rather than muskets; such was the desperation to hold the French back and prevent total collapse.

Russian cavalry was organized around the squadron. Cuirassier and dragoon regiments tended to have four or five squadrons; hussar and uhlan regiments were larger, with eight to ten squadrons. Cuirassiers were the heavy cavalry of the Russian army, armed with a straight sabre and a pair of pistols. At times they were also supplied with cavalry carbines or rifles. The cuirassiers were considered the elite troops and the best men from other cavalry regiments were drafted in to fill the ranks. In 1812 the most numerous Russian cavalry type was the dragoon, of which there were thirty-six regiments. However, by 1813 there were just eighteen left, many of them having become uhlans, horse jägers or cuirassiers.

The St Petersburg Dragoons were, perhaps, the most famous. Throughout the conflict they captured no less than four French colours. Russian horse jägers were broadly equivalent to the French chasseurs. Again they were light cavalry and eight dragoon regiments were converted to horse jägers in 1812. The troopers were given sabres, a pair of pistols and a carbine.

The Russians also deployed twelve uhlan regiments in 1812. They were primarily from western Russia and drawn from Poles and Lithuanians living in that part of the country. There was also a regiment that was composed entirely of Tartars. They were considered to be expert horsemen and they were armed with a sabre, a pair of pistols and a lance.

Russian hussars were mainly drawn from volunteers rather than pressed men. They were considered to be outstanding fighters, at least as good as any other hussar that could be fielded by another European nation. There were never enormous numbers of hussars, yet invariably they fought exceptionally well.

The Cossacks were incredibly important to the Russian army. They participated in all of the battles and campaigns and they were particularly effective during the French invasion of Russia in 1812. To the great horror of the French, Cossacks even reached Paris in 1814 and again in 1815. There were large numbers of Cossack regiments. Many were deployed in the border regions and in 1812 there were eighty-eight regiments of Don Cossacks alone and fifty or sixty other regiments from various parts of the Russian Empire. Invariably the Cossack carried an 8ft lance with a steel tip. They were all also armed with sabres and pistols. Others carried carbines and muskets.

The Life Guard Cossack Regiment was considered to be the best and the men were particularly selected for their strength and their height. The Cossacks were used to great effect on the flanks of the French army at Borodino and more generally they were able to spread terror along the supply lines, which was feeding and replenishing the French army deep into Russian territory.

A Don Cossack regiment numbered around 600 men and two or three regiments were formed into brigades. Some of the Cossack brigades were attached to regular troops, whereas others operated as separate corps.

At the end of the 18th century, the Russian artillery corps was in a fairly poor state. It did not really improve until after the Austerlitz campaign. The French captured a huge number of Russian guns and the need to rebuild was obvious. By 1812 the Russian artillery was far more professional, well trained, efficient and above all accurate in its fire.

Gunners tended to be trained in companies or batteries and officers were turned out by the 1st and 2nd Moscow Academy. In 1812 the Russians could field 176 12pdrs, 524 6pdrs, and the same number of foot'unicorns'(a mixture of 10 and 20pdrs), 132 6pdr horse guns and the same number of 10pdr unicorn horse guns. The guard artillery had a further sixty-four guns. The Russians

tended to deploy their batteries just forward of the infantry line, although they would be deployed in larger groups if they were in earthworks.

The Russians were the first to successfully use the grand battery. In 1807, at the Battle of Eylau, they used three batteries, one of seventy guns, another of sixty and the third with forty guns.

The Russians also had their own Imperial Guard. They were given the best uniforms, weapons and recruits. At Austerlitz in 1805 the Russians deployed a Guard infantry division and a Guard cavalry division, but this was considerably enlarged by 1812 where there was a full Guard infantry corps and a supporting cavalry corps. The Guard infantry originally consisted of 10 Life Guard Grenadier battalions and a further battalion of jägers. By 1811 there were five full Life Guard infantry regiments and a jäger regiment. This was further reinforced by 1813 and in addition to this there were other, smaller units, such as the Guard Marine battalion.

The Guard cavalry at the time of Austerlitz in 1805 consisted of a Guard cavalry regiment, a Life Guard horse regiment, Life Guard Hussars and Life Guard Cossacks. This too had been considerably reinforced and improved by 1814, where there were now seven Guard cavalry regiments, representing each different key type of Russian cavalry.

The Guard artillery was a relatively small force; in 1801 it had just fifty-three guns. By the time the Napoleonic Wars were over the Guard units went into steep decline and forty years later, during the Crimean War, they were more of a liability than an asset.

Chapter 3
Weapons and tactics

There have been a number of modern books, not to mention contemporary Napoleonic titles, which deal specifically with the weapons and the tactics of the period. This chapter aims only to highlight the main weapons that were used by the five key protagonists and to provide a broad introduction to the kind of tactics that were used.

Infantry tactics and combat

The primary weapon for the infantryman of the Napoleonic Wars was the smoothbore musket. The musket fired a spherical lead ball that was capable of inflicting dreadful damage. Infantrymen needed to use ready-made cartridges that contained both powder and ball in greased paper. Inside their cartridge pouches there were a number of slots, each containing some of these packages.

The process of loading and firing a musket was complicated and time consuming. The priming pan had to be opened, a cartridge taken from the pouch, ripped open and then some of the powder put into the priming pan. The priming pan was then closed. The rest of the powder was then poured down the musket barrel. A ramrod was then used to push the cartridge down onto the powder, with the cartridge paper acting as wadding. Only then could the infantryman cock the musket to fire.

Muskets would tend to foul after fifty or sixty shots. Weather had a detrimental effect on the cartridges and if the cartridges were allowed to get wet then they were useless. The rate of fire of infantrymen was somewhere between one and six rounds per minute. In practice it was usually around four. Target practice was very important, as every shot needed to count. The muskets in general were fairly accurate up to about 100 paces, but a reasonable shot could hit a man at 300 paces.

Conditions on the battlefield made enormous differences to the accuracy of the fire. When there was no wind an infantry unit would be engulfed by gun smoke, making it difficult to see anything. If the ball was not rammed home properly it could come out at an unpredictable angle. Misfires were common and perhaps as many as 1 in 5 shots were misfires. Infantry were not trained to necessarily aim at their targets, but fire in the general direction of the enemy. Equally, stress made firing effectiveness variable and more often than not men opened fire well beyond their effective range.

Types of firing depended on the nation. The French, for example, were trained to fire by company. They were also trained to fight by division (two companies), by half battalion or by battalion. When advancing they would fire by alternate companies. Each company would fire when they halted and the other companies would reload as they advanced.

The Prussians were taught to fire by half companies. The first rank would kneel and the second and third rank would stand and the battalion would fire from the right. The Prussians would also fire as they advanced. A half company would stop and fire then advance then the next half company would stop and fire.

The Russians tried to train their troops to fire at least three times a minute. The first and second ranks would fire first and then load and fire as quickly as possible. In fact one of the techniques used was to have the men in the third rank only load muskets. So they would be passed unloaded muskets and pass them back to the men in the second rank loaded.

Facing a regiment of around 600 men, an enemy could expect to be attracting as many as 3,000 shots per minute. Concentrated fire power was vital as smoothbore muskets were fairly inaccurate. The men would be closely packed, virtually elbow to elbow, and would be trying to load, aim and fire as quickly as they could. Generally, in a fire fight three ranks was the usual formation, with each rank firing in turn. There was the added danger that if the third rank was firing then the constant moving and jostling of the front two ranks could mean that one of the regiment's own men would be hit.

Fire fights could be extremely short and devastating affairs, or could last a considerable amount of time before the enemy showed any slackening of fire due to casualties. It took considerable resolve to stand and concentrate on loading, aiming and firing whilst men all around were falling. As soon as the situation changed and one side sensed the other was wavering, a bayonet attack, or at least the threat of one, would be enough to drive the enemy away.

The bayonet, effectively a dagger or knife that fitted over the muzzle of the musket, was a direct throwback to the pike. The first bayonets were of the plug variety; in effect the bayonet was pushed into the muzzle of the musket, which meant that the musket could not then be fired. Before the Napoleonic period, however, the socket bayonet had been designed. This slipped over the outside of the barrel, which meant that the musket could still be loaded and fired whilst the bayonet was attached.

The vast majority of casualties on a battlefield were not caused by bayonets during the Napoleonic period. In fact on average it was no more than around two per cent. Generally speaking, bayonet charges were only ever mounted against an enemy that had already been disorganized by fire, or was physically worn down. Most of the bayonet attacks therefore saw the target fleeing before the charge crashed into them. Bayonet fights tended to happen in difficult terrain, such as in built up areas or struggles to capture redoubts.

Grenadiers were considered to be the primary deliverers of bayonet attacks. Originally the grenadier was literally armed with a grenade as well as a musket, but by the Napoleonic Wars they were men that were used for close assault. They were taller and stronger and the idea was that they had more physical stamina and a height advantage in a bayonet fight.

Understandably, in the heat of battle, with smoke, noise and men being hit by potentially-unseen enemies, it was not unusual for men's nerve to be shattered. Added to this the unit might be under fire from artillery, literally bowling over ranks of men. Any retrograde move by part of the unit or by a regiment or battalion beside the unit could cause a panic amongst the men that could not be stopped. The loss of officers was serious and cohesion could easily be lost.

Moving around on the battlefield was, as far as infantry was concerned, a complex procedure and the formations would either depend on the circumstances or on the specific preferences and training of the nation. Before going into the different types of formation, the infantryman was either in close file, which meant that the men were virtually elbow to elbow,

which was ideal for forming up or creating a column, or they were in open files. This meant that they were more widely spread and in an ideal formation for skirmishing.

Each individual battalion tended to form up either in line, column, square or in open order for skirmishing. A regiment would form up, perhaps with two or three battalions in line. Then there would be a gap of up to 400 paces where the remaining battalions of the regiment would form up in column. This would give each battalion sufficient room to manoeuvre and adopt an alternative formation.

It was a perilous and potentially-chaotic prospect for an army to advance towards the enemy in line. Given the fact that the majority of the battlefields were not flat, unbroken plains but dotted with obstacles, by the time the army had advanced only a short distance the lines would be broken and ragged. The idea was to try to close to within effective range of the enemy in an efficient formation that gave a better guarantee that the unit would remain intact. This meant that columns, correctly spaced, would have to close with the enemy covered by skirmishers, but accepting the fact that casualties would be inevitable from enemy artillery or skirmishers. The battalions would have to be spaced so that at the chosen point they would transform from column into an unbroken line. Bear in mind that all of this had to be done under fire.

The British, for example, had a system where their lines would consist of three ranks, but in practice it was usually two (although at Waterloo they formed up in four ranks). British infantry went for a looser and more flexible system. The French was usually based on three ranks, but towards the end of the period Napoleon himself ordered that his infantry form up in two ranks, believing that the third rank was next to useless.

There were several types of column used by infantry but broadly they fell into two categories. There was the marching column, which was the preferred formation for travelling long distances on roads, or the column that was used on the battlefield. Sometimes these were used as attacking columns. The French favoured the column on the battlefield; it could move across even difficult terrain with relative ease and its speed meant that it would minimize the amount of time that the unit took to get into position. The problem was that the column was vulnerable to fire and that it was deficient in its own fire power because only the front ranks could actually fire with any effectiveness.

Mixed order was also used as a formation. Effectively this meant that it was a combination of lines and columns. The leading battalions would be in line, followed by battalions in column and then additional battalions would follow up in line.

In order to fend off cavalry, infantry formed squares. This was a complicated manoeuvre

on the battlefield and a dangerous one if cavalry was closing. On average, a unit would need up to five or six minutes to form a square, which meant that officers had to be on a constant lookout for cavalry threats. Most squares were hollow, although others such as the Prussians and the Austrians favoured a solid square. The size of squares could be enormous. At Krasne, in August 1812, the Russians created a square of sixteen battalions and fended off at least thirty charges by French cavalry.

If infantry had not formed up into square then they were a vulnerable target for cavalry, particularly if they were attacked from the flanks. On rare occasions lines or columns of infantry did withstand cavalry attacks but these were few and far between.

Skirmishers operated ahead of columns and lines. In theory all infantry was trained to operate as skirmishers. When skirmishers were lost more men would have to be detached from the parent unit to replace them. Skirmishers tended to focus on enemy officers, musicians, gunners and other skirmishers. They would fire off considerably more shots than a man standing in line or columns. Skirmishers, like other infantry, were vulnerable to attacks by cavalry.

French skirmishers, or *tirailleurs,* were trained to work in pairs. Only one man fired at a time so that the other always had a loaded weapon. The British also had their own specialist skirmishers, such as the 95th Rifles. Troops such as these were considered some of the best light troops in the world and are credited with having killed large numbers of French officers.

Cavalry tactics and combat

Heavy cavalry was used primarily as a shock force on the battlefield. They operated in tight formations and were often protected by armour. Their swords could outreach most of their opponents. The problem was that in a cavalry fight the heavier horse and the heavier man would find it more difficult to move around and would be vulnerable to lighter cavalry.

Generally, the light cavalry, such as lancers, uhlans and hussars, were the scouting forces of the army. They were designed to protect the army on the move, cover its flanks and be used to pursue or scout ahead. Most light cavalry was armed with shorter, curved sabres, whereas heavy cavalry would have the longer weapon. The vast majority of cavalry also had pistols, or carbines, or rifles. Other more specialist troops, like uhlans and lancers, obviously used the lance as their primary weapon. Most cavalry armed with swords tended to use them as a cutting and slashing weapon rather than a thrusting one and the vast majority of wounds and kills were on the right side of the enemy's body.

The basic tactical unit for cavalry was the squadron, effectively the mounted equivalent of the battalion. The speed at which the units moved depended on the quality of the horse and the amount of armour or equipment that the man was carrying. Generally speaking, at a walk cavalry would cover around 100 metres a minute. They would trot at double and gallop at around treble that speed.

Cavalry tended to form up either in squadron, regiment or brigade in lines, echelons or columns. The echelon was in effect an angled line with part of the line closer to the enemy and the rest of the line angling away from the enemy.

Whilst mounted, firing was complicated; aiming was not as easy from the saddle, so firing was usually individual or at best by squadron. Many cavalry were designed to be able to fight on foot, particularly the dragoons, whose roots were as mounted infantrymen. The mounted charge is seen by many to be the primary tactic of Napoleonic cavalry. The charge would begin at a slow pace, building up speed to the gallop. The slow pace allowed the formation to remain intact for as long as possible. If facing other cavalry it was usual for the enemy to countercharge rather than to stand still and receive the shock of impact, which could end in absolute disaster as the momentum would be with the charging unit. Whilst many cavalry charges were relatively small affairs, others could involve up to a hundred squadrons charging simultaneously.

Cavalry was especially useful in pursuit, when they could harry and continue to disorganize a retreating foe. When cavalry attacked squares they invariably lost out and were unable to break into the infantry formations. There was no particular pecking order in terms of effectiveness when cavalry faced one another. Lancers theoretically had the edge on sword-armed cavalry, but when lancers faced cuirassiers their points would invariably fail to penetrate the armour.

Artillery tactics and combat

During the Napoleonic period field artillery fell into two broad categories. There were the cannon, which were long-barrelled, low-trajectory weapons designed primarily for direct fire. Usually they would be 3, 4, 6, 8 or 12pdrs. There were also howitzers, which could fire at both high and low trajectories. They were indirect fire weapons.

Each gun was manoeuvred by a team of horses, usually up to eight of them. A caisson would be pulled by two horses. In most cases artillery operated in pairs and a battery would consist of between six and twelve guns. Batteries tended to be attached to infantry divisions, but some would be held back in reserve. In other cases grand batteries would be formed, where as many as 100 guns would be massed in order to fire at decisive points along the enemy lines.

The most common projectile fired by cannon was the solid ball, although they could also fire a cylinder that was filled with small balls, known as canister. Shells were also fired, which were explosives that burst into splinters. Shrapnel could also be used, which meant firing an iron case full of bullets, or sharp pieces of metal. Canister only tended to be used at close range, shell was used against barricades, woodland and buildings, and cannonballs were used against walls and targets at a longer distance.

After about five shots had been fired by a cannon, the barrel would usually become dangerously hot. It would have to be cooled down by using a wet sponge. Sponging the cannon after every shot was also important, as any sparks that remained in the tube could prematurely ignite the cartridge. Each time the gun fired it would jerk backwards, so the crew would have to push it forward again. Moving the guns around was also an onerous task and in muddy conditions more horses would be needed. Limbering up would take several minutes, while unlimbering was quicker but still took considerable precision and skills.

Much has been written about the range and accuracy of artillery fire. Firing at long range was never effective and if the gun had access to around 200 shots then it could easily fire all its ammunition off within two to three hours. Effective range for a 4pdr was no more than about 700 metres and only 900 metres for a 12pdr. 4pdrs could probably get off about two shots per minute and 12pdrs one shot per minute. For short periods of time, and particularly if the gunners were uncommonly skilful, the shots fired could be increased up to no more than around five to eight shots per minute.

Some calculations have been made about the effectiveness of artillery in particular battles. At the Battle of Wagram, and accepting the rough calculation that around 13% of Austrian casualties were caused by artillery, this means that one Austrian was killed for every 31 rounds fired by the French artillery, as they fired 96,000 rounds during the battle. Higher casualties could be expected if the artillery could make an oblique or flanking fire shot against formed infantry. Simply put, it would pass through more lines of enemy infantry than just slicing through a thin line.

On the battlefield most of the artillery units that were overrun were not taken from the front, but from the flank, as it would take a considerable amount of courage and determination to stare straight down the barrel of guns firing shrapnel at point-blank range.

Counter-battery fire was often considered to be a waste of ammunition, but it would be used if there were no other useful targets.

Austrians

General Alvintzi led a military reform commission in 1798 to look at the specific weapons and tactics used by the Austrian army. At the time the bulk of Austrian infantry was armed with a musket dating back to 1754, but Alvintzi wanted an M1798 pattern musket to be introduced. Obviously all of this would take time and the 1754 musket remained in common use throughout the Napoleonic Wars.

The German and Grenzer infantry regiments carried around sixty rounds of ammunition per man. They had a triangular bayonet and a 43cm-long fusilier sabre. The Hungarian and grenadier regiments were similarly armed, except they had a longer, 63cm grenadier sabre.

With the exception of the units that still retained their 1754 muskets, the bulk of the men were armed with the 4.2kg M1798 or the 4.6kg M1807. A carbine specifically for jägers, the M1807, weighed 3.55kg and a rifle, also an M1807, was introduced. Briefly, in the period 1797 to 1800, an M1780 repeating air rifle with a twenty-shot magazine was tried out by the jägers, but this proved to have major mechanical problems. Up until 1808 some Grenzers were armed with a short, double-barrelled weapon in addition to their rifles, but this was also later replaced.

Austrian infantry tended to use a combination of lines, columns, close order masses and skirmish chains as their primary tactics on the battlefield. It was generally considered that three-rank lines were the most effective for both attack and defence, although columns were certainly used. The columns would be screened by skirmishers. The Austrian infantry was trained to move at three speeds: 90, 105 and 120 steps per minute, but this would, of course, depend on the terrain. Some Austrian units also used to lie down when they were under artillery fire, which was designed to reduce casualties. When they were facing cavalry the infantry would adopt a battalion mass formation. In other words, it was a closely-packed column, just a company wide and six companies deep. It was considerably stronger than trying to face cavalry off in a line three ranks deep.

Generally speaking Austrian skirmishers were not that effective and in fact the Austrians were not great believers in the use of such troops. They recognized that they were useful in fending off enemy skirmishers, to mask their formed infantry, to cover withdrawals and to protect flank marches. Austrian skirmishers would tend to be drawn from men in the third rank of a line battalion, in addition to using jägers and Grenzers. They would take up a position around 300 paces in front of their battalion. On average, the Austrian battalion would push out up to 80 skirmishers.

Austrian advance guard units were almost exclusively light troops. They were fast, well drilled, made effective use of skirmishers and were certainly considered to be some of the better foes that faced the French.

Austrian cavalry were obviously issued with lighter weapons. The hussar carbine was 2.45kg, a heavier one at 3.25kg was used by dragoons and the cavalry rifle was 2.65kg. Austrian cuirassiers carried a large broadsword and a pair of pistols and early on each squadron had eight men with rifles and eight with carbines. Each squadron of dragoons had sixteen men that were armed with rifles, but the bulk of the men had a carbine and a heavy straight sword. Uhlan squadrons had eight men with rifles and eight with carbines. The bulk of the men had a pair of pistols, a curved sabre and a lance. The chevauleger were initially armed with a straight sword and carbine, but later a sabre and a carbine. Each squadron would also have sixteen men with rifles. The hussars carried a sabre and a hussar carbine.

To a large extent Austrian cavalry was not great when used as a solid mass. Comparatively speaking they had a fairly low proportion of cavalry to infantry, usually around one to seven, compared to the French, for example, of one to six.

When used with resolve, Austrian cavalry was capable of dealing with any reasonable threat. At Aspern-Essling in 1809 Lasalle's light cavalry was roughly dealt with by the Austrians, who charged both the front and flank of the French force. One of the major problems with the Austrian cavalry is that it did not seem to receive as much opportunity for training and exercises as the infantry and suffered on the battlefield as a result.

British

The British infantry was considered amongst the best in the world. Unlike many other nations, the British made use of specialist light infantry, which were drawn from the best marksmen available. There were, for example, eight battalions of the 60th Rifles and three battalions of the 95th.

The regular British infantryman was trained to march 70 steps per minute. This would be increased to 108 when moving in and out of column or line. The British combined two different approaches, which impacted on the type of formations they used and the tactics employed by their infantry. Their experiences in North America had taught them that it was important to be able to move around in open formations and use light infantry tactics, particularly in difficult terrain. But this was only truly effective if the enemy had very little artillery or cavalry. As an alternative, the British also continued the concept of using close order, disciplined drill and based most tactical formations on three ranks. Here we can see that the formations used in the Peninsular, for example, were markedly different from those used in other theatres. By

Waterloo, having fallen foul of French cavalry in the past, British infantry formed up in four ranks.

British cavalry were also considered to be superior to French cavalry man-for-man and in most engagements the British certainly came out on top. Wellington, however, had a fairly low opinion of his own cavalry and thought them almost incapable of having any tactical finesse on the battlefield. On occasion, however, British cavalry would come out with more than a bloody nose when they faced the French. At the Battle of Waterloo the Scots Greys were virtually wiped out by French lancers.

British artillery was efficient and well trained, but it was a relatively small part of the army. British guns were usually spread out along the battle line and were generally used to support the infantry. Only on a handful of occasions was the British artillery ever concentrated in order to create a break in the enemy line. At Waterloo, as the French cavalry surged towards them, the gunners quite rightly ran for the shelter of the infantry squares. As it transpired, few of the gunners found their way back to their guns and if Wellington had not created a reserve of artillery he would have had few guns capable of firing for the second half of the battle.

French

Most of the French infantry was armed with the Model 1777 Charleville musket. They also had a triangular bayonet and a short sabre. The musket was fairly accurate up to about 100 paces and a trained man could hit a target at up to around 300 paces.

Although the sabre was something of a burden, the French soldiers felt more capable of defending themselves when armed with it. When voltigeurs were ordered in 1807 to give up their sabres this was almost universally ignored, as was the similar edict to centre companies and chasseurs.

The organization of the French infantry underwent several changes. Before the Revolution regiments consisted of three battalions and in 1792 it was decided that instead of regiments, demi-brigades would be created. A demi-brigade consisted of a regular battalion and two volunteer battalions. By 1794 this had been adopted across the whole of the French army. Napoleon reinstated the regiment as the building block of the infantry in 1803. Demi-brigade remained as a term that was applied to provisional troops. Regiments would now have one depot battalion and between two and six battalions for war.

When Napoleon came to power each of the battalions would have one grenadier company and eight fusilier companies. In the period 1805 to 1808 the battalion consisted of one grenadier company, one voltigeur company and seven fusilier companies. After 1808 this changed again to one grenadier company, one voltigeur company and just four fusilier companies.

The backbone of the French infantry regiment was the fusiliers in line regiments, or chasseurs in light regiments. The muscle and elite force in each regiment were the grenadiers in line regiments or carabineers in light infantry regiments. Added to this were the voltigeurs, who were introduced by Napoleon himself. They were the smallest men from each company and were assigned to skirmishing duties.

In theory, the strength of a French cavalry regiment was between 800 and 1200 men. In practice, however, the strengths could be as little as thirty per cent of these totals. The building block of each regiment was the squadron, fielding between 85 and 250 men depending on circumstances. The most seasoned men were assigned to an elite company or the 1st Company of the 1st Squadron (although not in cuirassier or carabineers regiments).

In the aftermath of the Revolution the quality of the French cavalry had dropped, due to the large numbers of aristocrats that had left France for good. Napoleon was determined to make his cavalry as good as anything in the world. Whatever improvements were made by Napoleon in the period up to 1812, all of this was irretrievably lost by the enormous casualties of the Russian campaign. Many regiments literally ceased to exist. The cavalry had to be rebuilt almost from scratch in 1813. There were shortages of everything and particularly horses. The cavalry wing would never reach the pinnacle it had achieved prior to 1812. Even by 1813 standards, the cavalry was a shadow of its former self for the Waterloo campaign in 1815. Every regiment had been scaled back and only a short time prior to this many of the new cavalry had not ridden a horse before.

As soon as Napoleon came to power one of his major priorities was to bring root and branch change to the artillery of the French army. He helped establish new ways of producing ammunition, weapons and training. He also struggled to bring up the number of guns available to the army and improve the proportion relative to the infantry.

The French were not averse to using captured artillery pieces from other forces. The main difference between French artillery and that used by their opponents was that the French used their guns in an offensive manner. Their opponents tended to use it defensively to protect their infantry and cavalry. The allies would have a reserve of artillery, which would be allocated to divisional commanders on the battlefield in times of need. French artillery reserves, however, were kept back to deliver the final blow against an enemy and make a real difference to the outcome of the battle.

French artillery would be massed at a decisive point and would often mean the difference between success or failure on the battlefield. Like all other parts of the army, the artillery lost a huge amount of men, horses and equipment in the campaigns of 1812. Even the wood used to make the carriages and wheels after 1812 was of poorer quality because it was newly-cut timber and not seasoned. When the Waterloo campaign got underway Napoleon only really had time to rebuild his Guard artillery and could do little to improve the training of his regular gunners.

Prussians

The Prussian infantrymen were primarily armed with the Nothardt musket, which had been re-bored to a standard European calibre. From 1809 they used what was called the New Prussian musket, a far more effective weapon. Many Prussian infantrymen were also armed with British muskets, whilst others had French, Russian or even Swedish weapons. It was the Prussians that had first adopted the metal ramrod for their muskets, which meant that they could fight in three ranks, whilst Austrians before the Napoleonic Wars used wooden ramrods and in order to have the same rate of fire needed to form up in four ranks. In battle the Prussian infantrymen would invariably operate with their bayonets fixed to their muskets. Lighter troops would only put their bayonets onto their muskets when needed.

The Prussians used a number of different firing techniques until the infantry was reorganized in 1808. They would advance at speeds of between 75 and 108 paces per minute. By 1812 the Prussian brigade of two regiments would form up into three lines. They would retain this form with the nine-battalion brigade of 1815.

Prussian cuirassiers did not adopt body armour until 1814. The cuirassier was armed with a straight, heavy sword. Their dragoons had curved sabres and carbines and their uhlans had lances and curved sabres. Volunteer jägers attached to the uhlans would have a carbine instead of a lance. The carabineers had rifles and the hussars curved sabres and carbines.

Prior to 1806 the Prussian cavalry used a linear attack method. Effectively this was a two-rank line with a third rank posted on the wings of that line. The men on the wings could swing out and attack flanks and rears of enemy units if they had the opportunity. By 1812, Prussian cavalry was definitely seen as a subsidiary arm to the infantry. The cavalry was held back, deployed behind the infantry or on the flanks. They would only be unleashed if there was a definite opportunity to use them. This does not mean to say that Prussian cavalry was not aggressive or capable of handling themselves. When attacking infantry, particularly in square, they were ordered to attack the corners of the square.

The Prussians lost an enormous amount of artillery in the 1806 campaign and this meant that they had to largely rebuild their artillery forces, gradually increasing it in number and importance as the years passed. By 1813 they had 400 field pieces in 50 batteries, 38 of which were foot batteries.

Tactically Prussian artillery was supposed to engage enemy artillery by drawing fire on themselves. Their secondary objective was to fire at enemy columns and to prepare the way in general terms for the deployment of the Prussian army. The Prussians' heavier guns tended to concentrate on the enemy whilst they were deploying and at targets of opportunity, particularly columns. They would tend to do this at fairly long range. Prussian artillery was perfectly prepared to defend down to the last shot, even when the enemy was within touching distance.

There was a tendency, however, to keep a large amount of artillery in reserve and only to bring it out of the reserve to replace a battery that had used up its ammunition and powder. This led to the practice of deployed artillery using their ammunition up as quickly as possible so that they could be pulled back into the reserve.

Russians

Poor quality muskets were a major problem for the Russian infantry at the beginning of the Napoleonic Wars. Two different calibres of musket were in production between 1805 and 1809, although this went down to one calibre in 1809. In fact so confused was the situation that upwards of twenty-eight different calibres were in use in 1812.

The standard 1805 musket was heavy, at 5.16kg. New units and older units with worn out weapons were issued with the 1808 weapon, which weighed around 4.47kg. Upwards of 60,000 infantry were supplied with the British Brown Bess musket. Generally speaking, infantry was instructed to always operate with a fixed bayonet.

Rifles were also in production in Russia. In 1808, however, these were withdrawn from general issue and would henceforth only be used by marksmen in jäger companies. A major problem for the Russian infantry was the over-insistence of learning parade ground-style

precision in marching, as opposed to learning how to use their weapons. Target practice came fairly low down on the agenda.

As a general rule, an infantry division would form up in two lines. The first line comprised of the first infantry brigade. There would be around four battalions and they would adopt either a line or column formation. Up to 30 paces behind them was another infantry brigade, again of four battalions and in column. Ahead of the division was a partial jäger brigade operating as skirmishers. Russian infantry moved at four different paces. The slow speed was around 60 to 70 paces per minute. They then could break into a jog of around 100 paces per minute and a rapid run or charge of around 140 paces per minute. Infantry could also move at a rather more rapid pace in order to keep up with faster moving skirmishes.

The men were trained to fire volleys by platoons and by battalions. This meant that aimed fire was not exactly the forte of the Russian infantryman. In any case, Russian powder was of poor quality. There was also a shortage of bullets and a huge number of the muskets were worn out. Russian infantrymen therefore found themselves at an enormous disadvantage when they faced equivalent numbers of French infantry.

The Russians also used a far deeper type of column than an equivalent French battalion. This made it a much more tempting target for enemy artillery. Russian infantry were discouraged from using cover or from lying down to avoid artillery fire. Ahead of the main body were often insufficient numbers of jägers. This meant that the Russian battalion manoeuvring in a tight-packed formation around the battlefield was extremely vulnerable.

Russian cavalry had a wide variety of different edged weapons and firearms, as with any cavalry wing of the period. There were dragoon muskets, rifles, carbines and pistols. Some were home produced, others captured and some purchased from Britain or from Germany. Dragoons would carry a pair of pistols, a long musket and a bayonet. Musketoons were issued to some cuirassier and hussar regiments. These were lighter than infantry muskets and fired rather like a shotgun. Most hussars, however, were equipped with carbines, though some were allowed to keep their musketoons. Only the uhlans of the front rank were given lances in addition to a sabre and a pair of pistols. The second rank had carbines and the flankers, of which there were sixteen in each squadron, had rifles. Most cuirassiers had carbines and their flankers also had rifles.

The primary hand-to-hand weapon however was the sabre and they used a mix of curved and straight weapons. The sabre was used to cut with a slicing action. The heavier cavalry sabre was used by the cuirassiers and was a vicious weapon that was later replaced by a thinner weapon, similar to the one used by the French. Dragoons had a specific cavalry sabre. It had been copied from French dragoons and it allowed a thrust rather than a slash to be used.

The huge numbers of Cossacks in the Russian army had primarily an 8ft-long lance with a steel head. Many Cossacks also had sabres and any number of pistols. Some also carried muskets, carbines and other firearms. The lance was their main weapon, but they were perfectly prepared to use their sabres if necessary. The Cossacks were fearsome opposition for isolated infantry units. As far as the French were concerned, they were to be especially feared in skirmish situations. But they would not stand if charged by formed cavalry.

Before the Napoleonic Wars, Russian artillery was rather poor. It was used in an uncoordinated manner and not very accurate. They also carried with them very little ammunition. By 1812 it had improved markedly and by 1813 its fire power and accuracy were something to be admired.

Tactically, the Russians tended to use their artillery in powerful batteries of up to seventy guns. But it was invariably used to defend their army rather than being used as an offensive wing. The Russians also packed their battle lines with a number of guns, with the idea that it would make it more difficult to break through the line. The Russians seemed to be very attached to their guns and would only leave them at the very last minute; invariably they would try to save the weapon. Russian batteries tended to be deployed only just in front of the infantry lines and would look to the infantry to give them protection if the enemy got too close.

Chapter 4
Organizing troops

The following organizational charts aim to outline a typical division that represents a historical formation. They simply aim to reflect the balance of troops used at a particular point during the Napoleonic period. They are neither balanced forces in themselves, nor are they all equally powerful. Obviously multiples of these divisions or amalgams of divisions could be used. However, the structure usually follows the format of having a divisional commander then a number of brigade commanders. The allocation of commanders aims to reflect the specific difficulties in managing and commanding these forces.

Each type of commander has been given a nominal grading, although you may wish to change this to reflect particularly good or bad commanders from historical situations. Each of the units has also been given a nominal training and morale class. Where appropriate, units have also been identified as those that operate in extended or skirmish order as their primary formation on the battlefield.

Austrian Advanced Guard Division, 1805 (Austerlitz Campaign)

This is a typical advanced guard division, you may wish to combine all of the light guns from the 1st Brigade into a single battery. You can also do the same for the 2nd Brigade.

General of the Division	Gifted
1st Brigade General	Able
1st Grenz Regiment	Experienced Steady
2nd Grenz Regiment	Experienced Steady
3rd Grenz Regiment	Experienced Steady
Hussar Regiment	Experienced Steady

Notes: each of the Grenz regiments had a single light battalion gun attached. For wargame purposes there are 3 light guns which may be used as a single unit or attached to each regiment. The 1st Grenz regiment may be a jäger regiment with rifles. The hussar regiment is counted as light cavalry.

2nd Brigade General	Able
1st Line Regiment	Veteran Steady
2nd Line Regiment	Experienced Steady
3rd Line Regiment	Experienced Steady
4th Line Regiment	Experienced Steady
5th Line Regiment	Experienced Steady
Chevauleger Regiment	Experienced Steady

Notes: each of the line regiments had a single light battalion gun attached. For wargame purposes there are 3 light guns which may be used as a single unit or attached to each regiment. The first line regiment are grenadiers. The chevaulegers are light cavalry.

Austrian Line Division, 1805

This is a typical Austrian line division based on two brigades with attached cavalry support. This is a good defensive division. You may combine the light guns from the 1st Brigade into a single large battery. You may also combine the light guns from the 2nd Brigade into a single battery.

General of the Division	Able
1st Brigade General	Able
1st Line Regiment	Veteran Confident
2nd Line Regiment	Trained Confident
3rd Line Regiment	Trained Confident
4th Line Regiment	Trained Confident
5th Line Regiment	Trained Confident

Notes: each of the line regiments had a single light battalion gun attached. For wargame purposes there are 3 light guns which may be used as a single unit or attached to each regiment. The 1st Line Regiment may be grenadiers, if so they are classed as Elite Confident.

2nd Brigade General	Able
1st Line Regiment	Veteran Steady
2nd Line Regiment	Experienced Steady
3rd Line Regiment	Experienced Steady
4th Line Regiment	Experienced Steady

Notes: each of the line regiments has a single light battalion gun attached. The 1st Line Regiment may be grenadiers, if so they are classed as Elite Confident.

Attached Chevaulegers	Experienced Steady

The chevaulegers are light cavalry.

Austrian Reserve Division, 1805

Although more difficult to control than the advance guard or the line division, this is a perfectly capable formation, particularly in defence. You may combine all of the light guns from the 1st Brigade into a single battery. You may also combine all of the light guns from the 2nd Brigade.

General of the Division	Cautious
1st Brigade General	Cautious
1st Grenadier Regiment	Experienced Steady
2nd Grenadier Regiment	Experienced Steady
3rd Grenadier Regiment	Experienced Steady
4th Grenadier Regiment	Experienced Steady

Notes: each of the line regiments had a single light battalion gun attached. For wargame purposes there are 3 light guns which may be used as a single unit or attached to each regiment.

2nd Brigade General	Cautious
5th Grenadier Regiment	Experienced Steady
6th Grenadier Regiment	Experienced Steady
7th Grenadier Regiment	Experienced Steady

Notes: each of the line regiments had a single light battalion gun attached. For wargame purposes there are 3 light guns which may be used as a single unit or attached to each regiment.

Attached Cuirassiers Regiment	Experienced Steady
Attached Dragoon Regiment	Experienced Steady

The cuirassiers are heavy cavalry

Austrian Advanced Guard Division, 1809

An advanced guard division would be present in each Austrian corps. The medium artillery battery can be split and 2 guns attached to each brigade.

General of the Division	Gifted
1st Brigade General	Gifted
1st Jäger Regiment	Experienced Steady
2nd Jäger Regiment	Experienced Steady
3rd Jäger Regiment	Experienced Steady
4th Jäger Regiment	Experienced Steady
Uhlan Regiment	Experienced Impetuous

Notes: the 1st Jäger regiment is armed with rifles. The Uhlans are lancers.

2nd Brigade General	Gifted
1st Grenzer Regiment	Experienced Steady
2nd Grenzer Regiment	Trained Steady
Cheveauxleger Regiment	Experienced Steady

Notes: the 1st Grenzer regiment is armed with rifles. The cheveauxlegers are light cavalry.

Attached Foot Artillery 4 medium guns	Experienced Steady
Attached Horse Artillery 2 light guns	Experienced Steady

Austrian Regular Division, 1809

The organization of this division can be adapted in a number of ways. Firstly, the hussar regiment can be replaced with a two-gun 12pdr heavy foot battery. This is to reflect the fact that the light cavalry was invariably concentrated in the advanced guard division. The second alternative is that both brigades can be German instead of having one Hungarian. The division also lacks heavy cavalry and grenadiers as these were combined into the reserve division.

General of the Division	Able
1st Brigade General	Able
1st Line Regiment (German)	Experienced Confident
2nd Line Regiment (German)	Experienced Confident
3rd Line Regiment (German)	Trained Steady
4th Line Regiment (German)	Trained Steady
2nd Brigade General	Able
5th Line Regiment (Hungarian)	Trained Steady
6th Line Regiment (Hungarian)	Trained Steady
7th Line Regiment (Hungarian)	Trained Steady
8th Line Regiment (Hungarian)	Trained Steady
Attached 1st Foot Artillery 4 medium guns	Experienced Cautious
Attached 2nd Foot Artillery 4 medium guns	Experienced Cautious
Attached Hussar Regiment	Experienced Steady

Notes: each of the foot artillery batteries can be attached to each of the infantry brigades. The hussars are light cavalry.

Austrian Reserve Division, 1809

The better infantry units of the Austrian army were combined into the reserve division. This is a tough division with plenty of staying power. Reserve divisions fought extremely well at Aspern-Essling and pressed on despite heavy casualties. The division can also have a four-gun 12pdr heavy gun battery. This would allow this battery to be attached to one of the infantry brigades and the medium battery to the other brigade.

General of the Division	Cautious
1st Brigade General	Cautious
1st Line Regiment (German)	Veteran Confident
2nd Line Regiment (German)	Veteran Confident
3rd Line Regiment (German)	Veteran Confident
4th Line Regiment (German)	Veteran Confident
2nd Brigade General	Cautious
5th Line Regiment (Hungarian)	Veteran Steady
6th Line Regiment (Hungarian)	Veteran Steady
7th Line Regiment (Hungarian)	Veteran Steady
8th Line Regiment (Hungarian)	Veteran Steady
Attached Foot Artillery 4 medium guns	Novice Steady
Attached Cuirassier Regiment	Experienced Steady

Notes: the foot artillery battery can be split into two 2-gun batteries and attached to each of the infantry brigades. The cuirassiers are heavy cavalry.

Austrian, 1809 (in the Tyrol)

The Austrians encouraged the rebellion in the Tyrol region. This Austrian force is typical of the troops sent to support the rebels against the Bavarians, Saxons, Westphalians and troops from the Rhine Confederation. They would also face French troops. It would take considerable time and casualties to suppress the rebellion. Many of the actions involved ambushes or surprise attacks against the Bavarians and others.

General of the Division	Gifted
1st Tyrolean Brigade General	Gifted
Schützenkompanie	Experienced Steady
1st Landschützen Regiment	Experienced Steady
2nd Landschützen Regiment	Experienced Steady
3rd Landschützen Regiment	Experienced Steady

Notes: the schützenkompanie are rifle-armed. Optionally, one of the landschützen regiments can be rifle-armed.

2nd Tyrolean Brigade General	Able
1st Landsturm Regiment	Trained Confident
2nd Landsturm Regiment	Trained Confident
3rd Landsturm Regiment	Trained Confident
4th Landsturm Regiment	Trained Confident
5th Landsturm Regiment	Trained Confident

Notes: these Landsturm regiments operated as light infantry and may be readily used in either extended lines or as skirmishers

3rd Tyrolean Brigade General	Able
4th Landschützen Regiment	Experienced Steady
5th Landschützen Regiment	Experienced Steady
6th Landsturm Regiment	Trained Confident

Notes: one of the landschützen regiments may be armed with rifles

Regular Brigade General	Able
Jäger Regiment	Veteran Confident
1st Hungarian Regiment	Trained Cautious
2nd Hungarian Regiment	Trained Cautious

Baden, 1809

This list is designed to represent the Baden contingent that fought with the French against the Austrians at Aspern-Essling in 1809. The list therefore includes both French and troops from Baden. At the time, Baden was part of the Confederation of the Rhine. The Baden contingent performed well at Wagram when it marched across the front of the Austrian army and then took up its station between the exposed left flank of the French army and the River Danube. The Baden contingent was a solid ally of the French until their country was overrun in 1813.

General of the Division	Gifted
French Brigade commander	Able
1st Battalion 23rd Line Regiment	Experienced Cautious
2nd Battalion 23rd Line Regiment	Experienced Cautious
1st Battalion 18th Line Regiment	Experienced Cautious
2nd Battalion 18th Line Regiment	Experienced Cautious
Attached French Foot Battery – 3 medium guns	Experienced Cautious

Notes: the 23rd Line Regiment can be deployed as light infantry.

Baden Brigade commander	Gifted
1st Battalion 1st Line Regiment	Experienced Steady
2nd Battalion 1st Line Regiment	Experienced Steady
1st Battalion 2nd Line Regiment	Veteran Confident
2nd Battalion 2nd Line Regiment	Veteran Confident
1st Battalion 3rd Line Regiment	Veteran Confident
2nd Battalion 3rd Line Regiment	Veteran Confident
Light Battalion	Veteran Confident

Notes: The light battalion are light infantry.

Attached cavalry commander	Able
Baden Chevauleger Regiment	Experienced Steady
Hesse Darmstadt Chevauleger Regiment	Elite Confident

Notes: both of the chevauleger regiments are light cavalry

Bavarian, 1809

There were three divisions in Lefebvre's French VII Corps. Bavarian troops fought during the Tyrolean Revolt and were highly successful in the Battle of Wagram (Wrede's Division). The organisation of this division would also be suitable for engagements in 1812.

General of the Division	Able
1st Brigade commander	Able
1st Leib Regiment	Veteran Steady
2nd Leib Regiment	Veteran Steady
1st Kronprinz Regiment	Experienced Steady
2nd Kronprinz Regiment	Experienced Steady
Light Battalion	Veteran Steady
Attached Foot Artillery – 2 medium guns and one light gun	Experienced Steady

Notes: the light battalion is armed with rifles.

2nd Brigade commander	Able
1st Salern Regiment	Trained Steady
2nd Salern Regiment	Trained Steady
1st Herzog Pius Regiment	Trained Steady
2nd Herzog Pius Regiment	Trained Steady
Light Battalion	Veteran Steady

Notes: the light battalion is armed with rifles

Attached Cavalry commander	Able
Dragoon Regiment	Veteran Steady
Chevauleger Regiment	Veteran Steady
Horse Artillery – 2 medium guns and 1 light gun	Veteran Steady

Notes: the dragoon regiment is heavy and the chevaulegers are light cavalry

Bavarian, 1813-1814

Technically, this is an Austro-Bavarian Division and it fought at Leipzig and later in France in 1814. The 5th Austro-Bavarian Corps was very badly led by Wrede at both Hanau and Leipzig, hence the poor grade allotted to the divisional commander. The Bavarians suffered enormous casualties in the war of 1812 against the Russians. In the aftermath of this, the army was reorganized with National Guard units to fill in the gaps left by the lost regiments.

General of the Division	Cautious
1st Brigade commander	Able
1st Chevauxleger Regiment	Novice Cautious
2nd Chevauxleger Regiment	Novice Cautious
3rd Chevauxleger Regiment	Novice Cautious
1st Line Infantry Regiment	Novice Cautious
2nd Line Infantry Regiment	Novice Cautious

Notes: The Chevauxleger Regiments are light cavalry

2nd Brigade Commander	Able
3rd Line Infantry Regiment	Raw Cautious
1st National Guard Regiment	Raw Cautious
1st Light Infantry Regiment	Trained Cautious
3rd Brigade Commander	Able
4th Line Regiment	Trained Cautious
5th Line Regiment	Raw Cautious
2nd National Guard Regiment	Raw Cautious
3rd National Guard Regiment	Raw Cautious
2nd Light Infantry Regiment	Trained Cautious
Attached Foot Artillery Battery – 2 medium guns, 1 medium howitzer	Experienced Steady
Attached Horse Artillery Battery – 2 medium guns, 1 medium howitzer	Experienced Steady

British in Portugal, 1808

This is the early force that was used by Wellesley in the Peninsular when he won his first battle against the French. Note that it lacks cavalry and this is to some extent made up by the fact that there are elite light units. It can be used in conjunction with a Portuguese division or enlarged to represent the eight brigades that were actually available at the time.

General of the Division	Gifted
1st Brigade commander	Gifted
5th Foot	Veteran Confident
9th Foot	Veteran Confident
38th Foot	Veteran Confident
2nd Brigade Commander	Gifted
36th Foot	Veteran Confident
40th Foot	Veteran Confident
71st Foot	Veteran Confident
3rd Brigade Commander	Gifted
5th/60th Rifles	Elite Confident
2nd/95th Rifles	Elite Confident
Attached Foot Artillery – 2 medium guns and 1 medium howitzer	Veteran Confident

Notes: The Rifle units are obviously armed with rifles and are considered light infantry

British in Portugal, 1808-1809

This force aims to represent a British force that faced the French when they invaded Portugal. It can be used in conjunction with a Portuguese brigade. Unlike the 1808 list it now has cavalry.

General of the Division	Able
1st Brigade commander	Able
1st Line Regiment	Veteran Confident
2nd Line Regiment	Veteran Confident
3rd Line Regiment	Veteran Confident
60th or 90th Rifles	Elite Confident

Notes: The light infantry are rifle-armed and can operate in extended or skirmish order

2nd Brigade Commander	Able
4th Line Regiment	Veteran Confident
5th Line Regiment	Veteran Confident
6th Line Regiment	Veteran Confident
Light Regiment	Elite Confident

Notes: The light regiment are not rifle-armed but they can operate in extended or skirmish order

Cavalry Brigade Commander	Able
Light Dragoon Regiment	Experienced Steady
King's German Legion Hussars	Veteran Confident
Attached Foot Battery – 2 medium guns and 1 medium howitzer	Veteran Confident

British, 1813-1814
This is the type of force that would have fought in the Peninsular War during this period and have also operated in the South of France. It was involved in the campaign in the Pyrenees and the invasion of southern France.

General of the Division	Gifted
1st Brigade commander	Able
1st Line Regiment	Elite Confident
2nd Line Regiment	Veteran Confident
3rd Line Regiment	Veteran Confident
2nd Brigade Commander	Able
4th Line Regiment	Veteran Confident
5th Line Regiment	Veteran Confident
6th Line Regiment	Veteran Confident
Portuguese Brigade Commander	Able
1st Portuguese Regiment	Experienced Steady
2nd Portuguese Regiment	Experienced Steady
3rd Portuguese Regiment	Experienced Steady
Cacadore Regiment	Experienced Steady

Notes: The Cacadore Regiment is light and may be armed with rifles

Attached Cavalry Brigade Commander	Able
Hussar Regiment	Experienced Arrogant
Attached Foot Battery – 2 medium guns, 1 medium howitzer	Experienced Confident

Notes: 1 British Rifle regiment, graded as Elite Confident, should be split between the 2 British brigades to simulate them operating as independent rifle companies. They would be armed with rifles and could operate as light infantry skirmishers.

British, 1815 (Waterloo Campaign)

This is a mixture of both British and Hanoverian units and includes top class units, such as the 42nd Highlanders and the 95th Rifles.

General of the Division	Gifted
1st Brigade commander	Gifted
North Gloucestershires	Veteran Confident
Cornwall	Veteran Confident
Cameron Highlanders	Veteran Confident
95th Rifles	Elite Confident

Notes: The North Gloucestershires are the 28th, Cornwall the 32nd and the Cameron Highlanders the 79th. The 95th Rifles are rifle-armed light infantry.

2nd Brigade Commander	Gifted
Royal Scots	Veteran Confident
Royal Highlanders	Elite Confident
East Essex	Veteran Confident
Gordon Highlanders	Veteran Confident

Notes: The Royal Scots are the 1st, the Royal Highlanders are the 42nd, the East Essex are the 44th and the Gordon Highlanders are the 92nd.

Attached British Foot Battery – 2 medium guns and 1 medium howitzer	Veteran Confident
Hanoverian Brigade Commander	Able
1st Landwehr Regiment	Raw Cautious
2nd Landwehr Regiment	Raw Cautious
3rd Landwehr Regiment	Raw Cautious
4th Landwehr Regiment	Raw Cautious
Hanoverian Foot Battery – 2 medium guns	Experienced Steady

Brunswick 1809

This is the force commanded by the Duke of Brunswick that went on to fight alongside the British during the Peninsular War and during the Waterloo Campaign. This force fought its way to the North Sea coast, where it was evacuated by the British. It is therefore a smaller than usual division and actually amounted to no more than around 2,000 men.

General of the Division	Gifted
Infantry Brigade commander	Gifted
Jager Regiment	Veteran Confident
Sharpshooters	Veteran Confident
1st Light Regiment	Veteran Confident
2nd Light Regiment	Veteran Confident

Notes: Both the Jagers and the sharpshooters are armed with rifles. These and the light regiments can operate as light infantry.

Cavalry Brigade Commander	Able
Hussar Regiment	Veteran Confident

Notes: The hussar regiment was actually a partial Uhlan regiment so half of the men could be armed with lances.

Attached Horse Artillery Battery – 2 medium guns	Experienced Steady

Brunswick, 1815 (Waterloo Campaign)

After the Duchy of Brunswick was liberated in 181,3 the Duke of Brunswick rebuilt his army and this division represents the force that fought alongside the British in the Waterloo Campaign.

General of the Division	Gifted
Avante Garde Brigade Commander	Gifted
Avante Garde Regiment	Veteran Confident
Jäger Regiment	Veteran Confident
Uhlan Regiment	Veteran Confident
Hussar Regiment	Veteran Confident

Notes: The Avante Garde are light infantry, the jägers are rifle-armed, the uhlans are lancers and the hussar regiment is classed as light cavalry

Light Brigade Commander	Able
Leib Regiment	Veteran Confident

1st Light Infantry Regiment	Novice Steady
2nd Light Infantry Regiment	Veteran Confident
3rd Light Infantry Regiment	Novice Steady
Line Brigade Commander	Able
1st Line Regiment	Novice Steady
2nd Line Regiment	Veteran Confident
3rd Line Regiment	Novice Steady
Foot Artillery Battery – 4 medium guns	Experienced Steady

Danish, 1813

This force represents the Danes that fought as French allies. It is a cut-down version of the army as there was actually another brigade with five more line regiments, dragoons and additional artillery.

General of the Division	Able
Avante Garde Brigade Commander	Able
Schleswig Chasseurs	Experienced Steady
1st Holstein Tirailleurs	Trained Cautious
2nd Holstein Tirailleurs	Trained Cautious
Jutland Hussars	Veteran Steady

Notes: To reflect the availability of rifles in this brigade the Chasseurs regiment could be rifle-armed. The Hussars are light cavalry

Infantry Brigade Commander	Able
1st Oldenburg Regiment	Novice Steady
2nd Oldenburg Regiment	Novice Steady
3rd Oldenburg Regiment	Novice Cautious
4th Oldenburg Regiment	Novice Cautious
Queen's Regiment	Experienced Steady
Holstein Dragoons	Novice Steady

Notes: The Dragoons are classed as light cavalry.

Attached Horse Artillery Battery – 2 medium guns, 1 medium howitzer	Experienced Cautious
Attached Foot Battery – 3 medium guns, 1 medium howitzer	Experienced Cautious

Duchy of Warsaw, 1813

This is a force that fought on the side of the French and saw action in Germany and later in Russia. It is an interesting force as all of the Polish infantry regiments are light infantry.

General of the Division	Able
1st Infantry Brigade Commander	Able
1st Infantry Regiment	Novice Steady
2nd Infantry Regiment	Novice Steady
3rd Infantry Regiment	Novice Steady

4th Infantry Regiment	Novice Steady
2nd Infantry Brigade Commander	Able
5th Infantry Regiment	Novice Steady
6th Infantry Regiment	Novice Steady
7th Infantry Regiment	Novice Steady
Cavalry Brigade Commander	Able
Uhlan Regiment	Novice Steady
Chasseurs Regiment	Novice Steady

Notes: The Uhlans are lancers and the Chasseurs are light cavalry.

Attached Foot Battery – 2 medium guns, 1 medium howitzer	Novice Steady
Attached Horse Battery – 2 medium guns, 1 medium howitzer	

Dutch Belgians, 1815

This is the force that fought alongside the British during the Waterloo Campaign. In 1814 they had actually been part of the French army. The list consists of both Dutch and Belgian troops, as well as those from Nassau who fought at Hougoumont during the Battle of Waterloo.

General of the Division	Able
1st Infantry Brigade Commander	Able
27th Dutch Jagers	Experienced Cautious
7th Belgian Line Regiment	Novice Cautious
5th Dutch Militia	Raw Cautious
7th Dutch Militia	Raw Cautious
8th Dutch Militia	Raw Cautious
Attached Foot Battery – 3 medium guns, 1 medium howitzer	Experienced Steady

Notes: The Dutch Jager Battalion are rifle-armed.

2nd Infantry Brigade Commander	Able
1st Nassau	Experienced Steady
2nd Nassau	Experienced Steady
3rd Nassau	Experienced Steady
1st Orange	Experienced Steady
2nd Orange	Experienced Steady
Nassau Jagers	Experienced Steady

Notes: The Jagers are rifle-armed

Attached Horse Battery – 3 medium guns, 1 medium howitzer	Experienced Steady

French, 1805

This was the French force that fought at Austerlitz and a reflection of the troops that would have been engaged in the invasion of Britain had it taken place.

General of the Division	Gifted
1st Brigade commander	Able
1st Legere Regiment	Experienced Steady
2nd Legere Regiment	Experienced Steady
3rd Legere Regiment	Experienced Steady

Notes: These regiments are considered light infantry.

2nd Brigade Commander	Able
1st Line Regiment	Experienced Confident
2nd Line Regiment	Experienced Confident
3rd Line Regiment	Experienced Confident
4th Line Regiment	Experienced Confident
5th Line Regiment	Experienced Confident
Cavalry Brigade Commander	Able
Hussar Regiment	Experienced Confident
Chasseur a Cheval Regiment	Experienced Steady
Attached Foot Battery – 3 medium guns, 1 medium howitzer	Experienced Steady

French in Portugal, 1808

This list is designed to match the British in Portugal 1808 and incorporates elite Swiss units. The complete army also had an additional infantry brigade and more dragoons.

General of the Division	Able
1st Infantry Brigade commander	Able
1st Legere Regiment	Elite Confident
2nd Legere Regiment	Experienced Steady
1st Line Regiment	Experienced Steady
2nd Line Regiment	Experienced Steady

Notes: The Legere Regiments are classed as light infantry and in practice the line regiments were all capable of skirmishing.

2nd Infantry Brigade Commander	Able
3rd Line Regiment	Veteran Confident
4th Line Regiment	Veteran Confident
Swiss Regiment	Elite Arrogant

Notes: In practice the Swiss Regiment was capable of skirmishing

Reserve Brigade Commander	Able
1st Combined Grenadier Regiment	Elite Arrogant
2nd Combined Grenadier Regiment	Elite Arrogant
Cavalry Brigade Commander	Able
Chasseur Regiment	Trained Impetuous
Dragoon Regiment	Trained Impetuous
Foot Artillery Battery – 3 medium guns, 1 medium Howitzer	Experienced Steady

French, 1809

This is a mix of raw and veteran troops that fought against the Austrians. It is a typical line division of the period.

General of the Division	Gifted
1st Brigade commander	Able
1st Legere Regiment	Experienced Steady
2nd Legere Regiment	Experienced Steady
3rd Legere Regiment	Experienced Steady
1st Line Regiment	Novice Steady
2nd Line Regiment	Novice Steady
3rd Line Regiment	Novice Steady

Notes: The Legere Regiments are light infantry

2nd Brigade Commander	Able
4th Line Regiment	Veteran Confident
5th Line Regiment	Veteran Confident
6th Line Regiment	Experienced Steady
7th Line Regiment	Experienced Steady
8th Line Regiment	Novice Cautious
9th Line Regiment	Novice Cautious
Cavalry Brigade Commander	Able
Hussar Regiment	Experienced Steady
Chasseur a Cheval Regiment	Novice Steady

Notes: Both of these regiments are considered to be light cavalry

Foot Battery Regiment – 3 medium guns, 1 medium howitzer	Experienced Steady

French in the Peninsular War to 1812

This is the force that fought against the Duke of Wellington and the Spaniards. In practice many of the regiments operated as skirmishers due to the nature of the warfare. The list lacks additional hussars and chasseurs that may also have been present at times.

General of the Division	Gifted
1st Brigade commander	Able
1st Legere Regiment	Experienced Steady
1st Line Regiment	Experienced Steady
2nd Line Regiment	Experienced Steady
3rd Line Regiment	Novice Steady
4th Line Regiment	Novice Steady
5th Line Regiment	Novice Steady

Notes: The Legere Regiment is light infantry and the two better line regiments can operate in skirmish order.

2nd Brigade Commander	Able
2nd Legere Regiment	Experienced Steady
3rd Legere Regiment	Experienced Steady
6th Line Regiment	Experienced Steady
7th Line Regiment	Experienced Steady
8th Line Regiment	Novice Steady
9th Line Regiment	Novice Steady

Notes: The two Legere regiments are light infantry and the two better line regiments can operate in skirmish order.

Cavalry Brigade Commander	Able
1st Dragoon Regiment	Experienced Steady
2nd Dragoon Regiment	Experienced Steady
Field Artillery Battery – 3 medium guns, 1 medium howitzer	Experienced Steady

French, 1814

This is a mixed French force that fought against a number of allied armies. It reflects the types of troops that were available to defend France at the time. Therefore there is a wide difference in quality.

General of the Division	Able
1st Brigade commander	Able
1st Line Infantry Regiment	Experienced Steady
2nd Line Infantry Regiment	Experienced Steady
3rd Line Infantry Regiment	Experienced Steady
4th Line Infantry Regiment	Novice Cautious
5th Line Infantry Regiment	Novice Cautious
6th Line Infantry Regiment	Raw Cautious
2nd Infantry Brigade Commander	Able
1st National Guard Regiment	Novice Cautious
2nd National Guard Regiment	Novice Cautious
3rd National Guard Regiment	Novice Cautious
4th National Guard Regiment	Novice Cautious
5th National Guard Regiment	Raw Cautious
6th National Guard Regiment	Raw Cautious
3rd Infantry Brigade Commander	Able
1st Naval Infantry Regiment	Novice Cautious
2nd Naval Infantry Regiment	Novice Cautious
3rd Naval Infantry Regiment	Raw Cautious
1st Provisional Regiment	Novice Cautious
2nd Provisional Regiment	Novice Cautious
Cavalry Brigade Commander	Able
1st Chasseur a Cheval Regiment	Novice Cautious
2nd Chasseur a Cheval Regiment	Novice Cautious

Notes: These cavalry regiments are light cavalry

Foot Artillery Battery – 3 medium guns, 1 medium howitzer	Experienced Steady

French, 1815 (Waterloo Campaign)

This is a typical rather than guard division that fought during the Hundred Days campaign and specifically during the Waterloo campaign.

General of the Division	Able
1st Brigade commander	Able
1st Legere Regiment	Experienced Steady
2nd Legere Regiment	Experienced Steady
3rd Legere Regiment	Experienced Steady
1st Line Regiment	Experienced Steady
2nd Line Regiment	Experienced Steady
3rd Line Regiment	Experienced Steady

Notes: The Legere regiments can operate as light infantry.

2nd Brigade Commander	Able
4th Line Regiment	Novice Cautious
5th Line Regiment	Novice Cautious
6th Line Regiment	Novice Cautious
7th Line Regiment	Novice Cautious
Cavalry Brigade Commander	Able
Hussar Regiment	Experienced Steady
Chasseur a Cheval Regiment	Novice Steady

Notes: Both of these regiments are considered light cavalry

Foot Artillery Battery – 3 medium guns, 1 medium howitzer	Experienced Steady

Hanover or King's German Legion, 1815 (Waterloo Campaign)

A considerable number of the infantry that fought under Wellington during the Waterloo campaign were in fact Hanoverian. They fought extremely well. It can be combined with a standard British infantry brigade and have additional foot artillery in order to more accurately represent precisely what was available during the campaign.

General of the Division	Gifted
King's German Legion Brigade Commander	Gifted
1st Light Regiment	Elite Confident
2nd Light Regiment	Elite Confident
5th Line Regiment	Elite Confident
8th Line Regiment	Elite Confident

Notes: Both of the light regiments are light infantry armed with rifles.

Hanoverian Brigade Commander	Able
Field Jager Infantry	Experienced Confident
Luneberg Regiment	Experienced Confident
Grunenhagen Regiment	Experienced Confident
Bremen Regiment	Experienced Confident
Verden Regiment	Experienced Confident

Notes: The Field Jagers are rifle-armed and the Luneberg and Grunenhagen regiments are light infantry.

Hanoverian Cavalry Brigade Commander	Gifted
Prinz Regent Hussar Regiment	Novice Steady
Bremen Verden Hussars	Novice Steady
Cumberland Hussars	Raw Impetuous

Notes: All of the cavalry regiments are light cavalry.

Foot Artillery Battery – 2 medium guns, 1 medium howitzer	Experienced Steady

Hesse-Darmstadt, 1809

This is essentially a Confederation of the Rhine force stiffened with a French brigade. Some of the units served in Spain, whilst the majority fought against the Austrians in 1809. They would also later see service in the Russian campaign of 1812.

General of the Division	Able
French Brigade commander	Able
1st Legere Regiment	Experienced Steady
2nd Legere Regiment	Experienced Steady
1st Line Regiment	Novice Steady
2nd Line Regiment	Novice Steady

Notes: The two Legere regiments are light infantry.

Hesse-Darmstadt Brigade Commander	Able
1st Leib Regiment	Experienced Steady
2nd Leib Regiment	Experienced Steady

1st Line Regiment	Experienced Steady
2nd Line Regiment	Experienced Steady
3rd Line Regiment	Experienced Steady
4th Line Regiment	Experienced Steady
Cavalry Brigade Commander	Able
Baden Chevauleger Regiment	Veteran Confident
Hesse-Darmstadt Chevauleger Regiment	Veteran Arrogant

Notes: Both of these regiments are considered light cavalry.

French Foot Battery – 3 medium guns, 1 medium howitzer	Experienced Steady

Kingdom of Holland, 1809

This is a force that was established after the French had invaded the country and it became the Kingdom of Holland. The regiments fought alongside the French army and most of them were lost during the 1812 campaign.

General of the Division	Able
1st Brigade Commander	Able
1st Line Regiment	Novice Steady
2nd Line Regiment	Novice Steady
3rd Line Regiment	Novice Steady
4th Line Regiment	Novice Steady
2nd Brigade Commander	Able
5th Line Regiment	Novice Steady
6th Line Regiment	Novice Steady
7th Line Regiment	Novice Steady
8th Line Regiment	Novice Steady
Cavalry Brigade Commander	Able
Cuirassier Regiment	Experienced Steady
Guard Hussar Regiment	Veteran Confident

Notes: The Cuirassier Regiment is heavy cavalry and the Hussars are light cavalry

Foot Artillery Battery – 3 medium guns, 1 medium howitzer	Experienced Steady
Horse Artillery Battery – 3 medium guns, 1 medium howitzer	Experienced Steady

Italian, 1809

This Italian force represents Eugene's army that fought alongside the French army against the Austrians. It has large numbers of light infantry and adequate guns.

General of the Division	Able
1st Brigade commander	Able
1st Line Infantry Regiment	Experienced Steady

2nd Line Infantry Regiment	Experienced Steady
3rd Line Infantry Regiment	Novice Steady
Istrian Regiment	Experienced Steady

Notes: The Istrian Regiment is light infantry

2nd Brigade Commander	Able
1st Light Infantry Regiment	Experienced Steady
2nd Light Infantry Regiment	Experienced Steady
3rd Light Infantry Regiment	Experienced Steady
4th Light Infantry Regiment	Experienced Steady

Notes: This whole brigade comprises of light infantry

Cavalry Brigade Commander	Able
Dragoon Regiment	Veteran Arrogant
Chasseur a Cheval Regiment	Trained Steady

Notes: The Dragoons are heavy cavalry and the Chasseurs are light cavalry

Foot Artillery Battery – 3 medium guns, 1 medium howitzer	Novice Cautious
Horse Artillery Battery – 2 light guns	Experienced Steady

Italians and Neapolitans in Spain, 1812

This is a force that was deployed in the Peninsula in 1812. They were involved in a number of anti-guerrilla operations.

General of the Division	Able
1st Brigade commander	Able
1st Legere Regiment	Experienced Steady
2nd Legere Regiment	Experienced Steady
1st Line Regiment	Experienced Steady
2nd Line Regiment	Experienced Steady

Notes: The two Legere regiments are light infantry units

2nd Brigade Commander	Able
3rd Line Regiment	Novice Cautious
4th Line Regiment	Novice Cautious
5th Line Regiment	Novice Cautious
6th Line Regiment	Novice Cautious
Neapolitan Brigade Commander	Able
1st Legere Regiment	Raw Cautious
Du Roi Regiment	Experienced Confident
De la Reine Regiment	Experienced Confident
Chasseur a Cheval Regiment	Raw Cautious

Notes: The Legere Regiment and the chasseurs are light troops.

French Cavalry Brigade Commander	Able
4th Hussar Regiment	Experienced Steady
13th Cuirassier Regiment	Veteran Confident

Notes: The hussars are light troops and the cuirassiers are heavy cavalry

oot Artillery Battery – 3 medium guns, 1 medium howitzer	Experienced Steady

Ottoman Empire, 1806-1810

The majority of these troops were levies but they had a large number of cavalry in each division. Generally the Turks fought defensive battles. This division can be used for any of the conflicts against the Russians during this period.

General of the Division	Cautious
1st Janissary Regiment	Experienced Steady
2nd Janissary Regiment	Experienced Steady
Infantry commander	Able
1st Albanian Regiment	Raw Impetuous
2nd Albanian Regiment	Raw Impetuous
3rd Albanian Regiment	Raw Impetuous
Greek Arnauts	Novice Cautious
Serbian Regiment	Novice Cautious

Notes: All of the infantry can operate in skirmish order and prefer close-quarter combat rather than trading musket shots.

Cavalry Commander	Able
1st Spahi Regiment	Veteran Steady
2nd Spahi Regiment	Experienced Steady
Albanian Cavalry	Novice Cautious

Notes: The Albanian cavalry is capable of operating as skirmishers.

Foot Artillery Battery – 2 heavy guns, 2 medium guns and 1 light gun	Experienced Steady

Prussian, 1806

These were generally very poorly led divisions, but they had reasonable infantry and a large number of artillery. There were usually two brigades to each division and two brigades of cavalry. In all cases each of the infantry regiments should normally be given a light gun.

General of the Division	Poltroon
1st Brigade commander	Cautious
1st Grenadier Regiment	Veteran Steady
1st Line Regiment	Veteran Steady
2nd Line Regiment	Veteran Steady
3rd Line Regiment	Novice Steady
4th Line Regiment	Novice Steady
2nd Brigade Commander	Cautious
2nd Grenadier Regiment	Veteran Steady

5th Line Regiment	Veteran Steady
6th Line Regiment	Veteran Steady
7th Line Regiment	Novice Steady
8th Line Regiment	Novice Steady
Attached Cuirassiers	Veteran Steady

Notes: The Cuirassiers are heavy cavalry

Foot Artillery Battery – 3 heavy guns, 1 medium howitzer	Novice Steady

Prussian, 1813-1815

This is a very different look Prussian army. It was involved during the invasions of France and the Waterloo campaign. The three separate brigades represent the three different types of Prussian infantry.

General of the Division	Able
Jäger Regiment	Experienced Steady

Notes: The jägers are armed with rifles and may operate as light infantry.

Regular Brigade commander	Able
Fusilier Regiment	Experienced Steady
1st Musketeer Regiment	Experienced Steady
2nd Musketeer Regiment	Experienced Steady

Notes: The Fusiliers may operate as light infantry

Reserve Brigade Commander	Able
1st Line Regiment	Experienced Steady
2nd Line Regiment	Novice Steady
3rd Line Regiment	Novice Steady
Landwehr Brigade Commander	Able
1st Landwehr Regiment	Novice Steady
2nd Landwehr Regiment	Raw Steady
3rd Landwehr Regiment	Raw Steady
Cavalry Brigade Commander	Able
Hussar Regiment	Experienced Steady
Dragoon Regiment	Experienced Steady
Landwehr Uhlan Regiment	Novice Cautious

Notes: The hussars are light cavalry, the dragoons are heavy cavalry and the uhlans are lancers

Foot Artillery Battery – 3 medium guns, 1 medium howitzer	Experienced Steady

Russians in the Balkans, 1810

These are the Russian troops that fought against the Ottoman Turks in a war that had been going on since 1806. These units would also be recalled to Russia to fight in 1812.

General of the Division	Able
1st Brigade commander	Able
Fangoria Grenadiers	Veteran Arrogant
1st Vitebsk Musketeers	Experienced Steady
2nd Vitebsk Musketeers	Experienced Steady
Orlof Musketeers	Novice Steady
2nd Brigade Commander	Able
1st Novgorod Musketeers	Novice Steady
2nd Novgorod Musketeers	Novice Steady
1st/6th Jagers	Veteran Confident
2nd/6th Jagers	Veteran Confident

Notes: The Jagers are classed as light infantry

Cavalry Brigade Commander	Able
Smolensk Dragoons	Experienced Steady
Andrenov Cossacks	Novice Steady

Notes: The Dragoons are classed as heavy cavalry and the Cossacks are capable of skirmishing.

Light Artillery Battery – 2 medium guns, 2 medium howitzers	Experienced Steady

Russians from 1812

This is the force that fought in the War of 1812 and later in both Germany and France. It should have two line brigades, a jäger brigade and additional artillery and these can be added to give it a far broader scope.

General of the Division	Able
1st Brigade commander	Able
1st Musketeer Regiment	Experienced Steady
2nd Musketeer Regiment	Experienced Steady
3rd Musketeer Regiment	Novice Steady
4th Musketeer Regiment	Novice Steady
2nd Brigade Commander	Able
5th Musketeer Regiment	Experienced Steady
6th Musketeer Regiment	Novice Steady
7th Musketeer Regiment	Novice Steady
Jäger Regiment	Veteran Confident

Notes: The jägers operate as light infantry

Cavalry Brigade Commander	Able
Lancer Regiment	Experienced Steady
Hussar Regiment	Experienced Steady
Cossack Regiment	Novice Steady

Notes: The hussars are classed as light infantry and the Cossacks are capable of skirmishing.[on foot or mouted. Rules do not allow Cavalry skirmishers]

Light Artillery Battery – 4 medium guns, 2 medium howitzers	Experienced Steady

Spanish, 1811

This is essentially an irregular force that fought against the French during the Peninsular War. It contains some decent units, but the rest are rather unreliable.

General of the Division	Cautious
Vanguard Brigade commander	Cautious
Murcian Light Infantry	Novice Cautious
Canarias Regiment	Novice Cautious
Leon Regiment	Novice Cautious
2nd Brigade Commander	Able
Rea Regiment	Novice Cautious
Zamora Regiment	Novice Cautious
Voluntarios de Navarra Regiment	Novice Cautious
3rd Brigade Commander	Cautious
2nd Guards Regiment	Veteran Confident
Walloon Guards Regiment	Veteran Confident
Irlanda Regiment	Veteran Confident
Patria Regiment	Veteran Confident
Toledo Regiment	Novice Steady
Legionestranjera Regiment	Novice Steady
Cavalry Brigade Commander	Able
Santiago Regiment	Novice Steady
Castilla Hussar Regiment	Experienced Steady
Granderos Regiment	Veteran Confident
De Instrucion Regiment	Veteran Confident

Notes: All regiments are heavy cavalry except the hussars, which are light

Foot Artillery Battery – 3 medium guns	Novice Cautious

Chapter 5
Wargame Rules

These rules represent the next logical step in wargame rule development. They are designed for the beginner and the hardened campaigner alike; above all, we have designed the rule systems to take the hard work out of calculating cause and effect of actions and situations. A series of tables, which may appear daunting to begin with, are the core of the system. Once you have become confident in using these tables, the results will become apparent and memorable. All the charts use ten-sided dice (D10).

We have endeavoured to design a system unlike anything else you may have seen before. Certainly there will always be loopholes in even the most precise of rules; we urge you take adopt the spirit of the game and not to dwell on gamesmanship, but on the game itself. By playing the game for its entertainment merits, its historical accuracy and a very real desire to recreate the period itself, you will achieve and enjoy the essence of wargaming itself.

Scales and movement are designed primarily for 15, 20 and 25-28mm figures. For smaller scales wargamers might want to adapt the movement, ranges and other distance issues down to suit them. Using smaller-scale figures the movement and ranges may seem too long, but this will be very dependent on the size of table you want to use. Maintaining these distances will still give you a very good game and the advantage over larger scale figures, the open flank.

The rules do not have a points system, but they do have clear criteria for winning and losing battles. Rather than trying to create equal-sized forces, super-troopers and lopsided armies, we have opted for accurate scaled down versions of real regiments, battalions, brigades, divisions, corps and armies. To wargame the Napoleonic Wars in any other way would take away the very nature of the conflict. Be prepared to be outnumbered, be prepared to have the odds stacked against you. All the more satisfying when you achieve your objectives on the battlefield against a superior force.

There are four simple stages to set up and play each game turn, within each game turn there are set procedures and an order in which to do things. Remember everything is reliant upon your initial reactions and orders to the situation, you can change things, but this takes time. Study the battlefield, consider your enemy and issue appropriate orders taking account of the strength of your forces and what is reasonable for them to achieve. There is no helicopter view or all-seeing-eye, to react to changes you must follow the chain of command, just as Wellington or Napoleon would have done on the battlefields of Europe two hundred years ago.

1 Preparing the Game

1.1 Agree scenario
1.2 Set up table and allocate objectives and features
1.3 Organize troops into brigades, note unit training and experience
1.4 Determine commander ratings.
1.5 Prepare brigade cards and shuffle for first turn
1.6 Deploy.
1.7 Issue army orders down to brigade level

1.1 Agree Scenario

Three scenarios are included in this book. Many further scenarios will also be available in the accompanying Grand Battery scenario books. Of course you may also make up your own.

1.2 Set up table and allocate objectives and features.

If you are using one of the pre-prepared scenarios, simply set up your table according to the map provided (tailoring it as necessary to the scenic items available). Make sure both sides are clear about the designation of various features: are rivers fordable or unfordable? Are hills steep hills or shallow? Allocate objectives (see section XX XX Objectives and Winning)

1.3 Organizing your army

All units are organized along regimental, divisional, brigade or corps lines. Infantry regiments consist of 5 stands of figures. For gaming purposes you need not worry about how many figures you place on each base. We suggest 2 to 6 figures per base. For aesthetic purposes you might depict specialist skirmisher units, or duplicate figures for line battalions when skirmishing, with just 2 or 3 figures per base, so they look more dispersed and can be placed less regularly.

Cavalry regiments are 3 stands of figures, ideally with 2-3 figures per base.

Artillery is represented by single artillery models, one representing a battery. It is optional whether you paint and base limbers and horses to represent the artillery on the move. For the specifics of the nation of your choice, see the chapter organizing your troops.

Regiments and batteries are organized into brigades, each with a brigade commander. Brigades are organized into divisions and divisions into corps (only the largest games will require multiple corps). Artillery batteries can usually be allocated to brigades or combined as a grand battery under the commander-in-chief (C-in-C). The C-in-C simply means the highest level of command present. In a game with two divisions per side, the C-in-C will be a corps commander, if only one division per side, he would be the divisional commander. This command structure is vital to the game and is laid out for you in the prepared scenarios.

Basing

For newcomers to the period we recommend the following base sizes for the five key scales:

	Frontage(mm)			**Depth (mm)**		
scale:	28/20 mm	10/15 mm	6mm	28/20 mm	10/15 mm	6mm
Troop type:						
Infantry	60	40	20	40	30	20
Cavalry	80	60	30	60	40	30
Artillery (deployed)	50	40	30	60	50	30
Artillery (limbered)	50	30	20	30	15	15
Skirmishers	60	40	30	20	15	20
Dismounted cavalry	60	40	30	20	15	20

Wargamers who have already based their units need not rebase, provided both sides have a consistent basing system.

Unit training and experience

Every unit is rated in terms of training and experience. There are five grades:

Green: Trained with no combat experience, an unknown quantity

Trained: Some battlefield experience, trained but not tested in actual combat

Experienced: Some combat experience, trained with an average ability and considered steady and reliable

Veteran: Trained with combat experience, more than one engagement, capable troops

Elite: Trained, steady, extensive combat experience, reliable/capable troops

Unit training has a big effect on actions during the game and must be determined before play starts. If you are using the ready-made scenarios, this has been done for you.

Unit morale

In addition to training and experience, each unit is also graded by its morale. Again there are five grades:

Impetuous
Cautious
Steady
Confident
Arrogant

If you are using the prepared scenarios, the morale of each unit is already determined. Otherwise you must decide and note it before play begins. Morale grades and their effects are covered in Section 8: Morale.

1.4 *Determine Commander Ratings:*

Just as the quality of units must be determined before play starts, so must the quality of the unit commanders. Again, these are prescribed in the ready-made scenarios but otherwise may be decided by mutual consent or generated randomly using the commander quality table (see Command and Control below).

1.5. *Prepare Brigade Cards:*

These are used to determine the order of play in each turn. You will need a card for each brigade commander, (and each division commander and each corps commander if the game is large enough) and one for the commander-in-chief of each side. If artillery batteries are allocated to brigades they will be covered by that brigade's card, and if massed in a grand battery they are covered by the commander-in-chief's card. Some scenarios may specify that you must also add a random event card in the pack.

The cards can be simply made from plain card with the name of the unit or commander written on one side (eg British 1st Brigade or French C-in-C). Alternatively, you can use ordinary playing cards, in which case make sure you can keep track of which card is allocated to which brigade (either a written list, or a system whereby, for example, the ace, two, three and four of hearts are the British first, second, third and fourth brigades etc). When you have created all the required cards, form these into a single deck, shuffle and place face down.

1.6 Deployment.

Once your table is set up and forces organized you may deploy. The ready-made scenarios will detail the restrictions on where each army may deploy. If you are not using a ready-made scenario, we suggest you use the brigade cards to determine the order of deployment randomly. If the first card drawn is, for example, the French commander-in-chief, he deploys his commander and any artillery concentrated as a grand battery, before the next card is drawn. Once the next card is drawn, the player may not go back and change deployments already made and must wait until the game begins to move those units (except by mutual consent with a friendly opponent of course!) Keep drawing cards until all units of each army are fully deployed. Don't forget to shuffle the cards again, ready for Turn 1.

In deploying it is worth noting that if you place a regiment or battery too far from its brigade commander it will immediately be 'out of command' (see Command Radius and Out of Command, below).

Players should agree a limit on how far forward from their own table edge each side may deploy. We suggest this should be a minimum of 9 inches from the centreline of the table, so the armies will be at least 18 inches apart and deploying safely out of range of rifles, though not artillery.

Permitted Formations

The prescribed formation for rapid movement in the Napoleonic Wars was the column. Regiments adopted the column for movement both along roads and across more difficult terrain. Columns were never intended to be the formation in which the regiments fought. At the earliest opportunity a regiment in column would deploy into line when faced by the enemy. An extended line might be used to hold a greater frontage, or troops dispatched to form a skirmishing screen. When threatened by cavalry, regiments would attempt to form square.

In order to recreate the period, we therefore have five permitted formations: column, line, extended line, square and skirmish order.

Column was primarily used for non-combat moves, where the regiment is in no immediate danger of contacting the enemy. This means that the column is an acceptable formation on the wargamers' tables, but that if immediately threatened by the enemy, the regiment will attempt to form a line to face that threat if it has the opportunity. The stands of the unit are placed adjacent, one behind the other in single file. Column may be used by infantry or cavalry.

Line is the most common formation. Stands are placed side by side in a single, unbroken line (fences, hedges and walls may prevent this, but it is still considered a line). The line is the standard formation for advancing, engaging and deploying. Line may be used by infantry or cavalry.

Extended line is an alternative to line formation, which represents the regiment adopting a looser formation, effectively occupying more frontage on the wargames table. This is a halfway house between line and skirmish order. An extended line consists of the regiment's stands in line

abreast, but with up to half the width of an individual stand between each pair of stands. Cavalry does not form extended lines.

Square affords infantry some protection from cavalry attacks. On the other hand, it makes the unit a more vulnerable target for musket and artillery fire. It is represented by stands facing outwards on all sides. To form 'square' from three bases, either arrange as a triangle or simply leave one face opem; with two bases, simply place them back to back. If a square falls to a single stand, it is no longer considered to be a square for movement, defence and target purposes. The owning player must immediately reform the unit into either a line or a column once this happens (see Section 4.3: Formation Changes).

To form a square from a full-strength unit of five stands, the command stand is placed in the centre of the square. Different units may form one square. However, stands of each unit in that square must still be in base to base contact and must not be jumbled up in different faces of the square amongst stands of other units.

Only infantry may form square.

Skirmish Order consists of infantry in very loose order. Skirmishers are able to move faster in all terrain compared to formed regiments. Their primary role is to provide a screen for the main body of troops. Only one regiment in a brigade can deploy as skirmishers at any one time, to do so the regiment forms a much-extended line. Each base must be a minimum of one base width apart from its neighbour. To act effectively as a skirmish screen for their main body, the skirmishers must be a minimum of 2 inches from the main body/brigade. A unit in skirmish order cannot be more than 6 inches from their main body.

Napoleonic armies did have regiments of dedicated skirmishers and marksmen, such as the British 95th Rifles. These types of units functioned as elite troops always in a skirmish role, and sometimes fought as detached companies, attached to infantry brigades. They were armed with rifles, giving them a better rate of fire than the ordinary infantry, and were superior marksmen, quite capable of sniping at officers or other important figures. However whilst so employed these two regiments were vulnerable to an infantry or cavalry attack, and would evade being drawn into close combat.

Dismounted cavalry operate in a very similar way to skirmishers, but they will not move further than 6 inches from their horse-holders.

Artillery may be deployed unlimbered and ready to fire, or limbered and ready to move.

1.7 Issue Initial Orders

Now that they have seen the relative dispositions of both armies, the players can formulate their plan and issue their initial orders down to brigade level and write these down. See the following section on command and control for an explanation of the types of orders permissible at each level of command and the restrictions they place on the orders given at the next level down.

2 Command and Control

Regimental commanding officers are not represented separately from their units; this reflects the limited influence these officers would have had on the course of the battle. However, here we look at the following:

Brigade commanders
Divisional Commanders
Corps commanders
Army commanders

Commander Quality

There are five grades of commander. Generic gradings are given in the lists in the chapter on Organizing Your Forces and are specified in the prepared scenarios (in this book or the accompanying scenario books). Alternatively they may be generated randomly using the following table:

Commander Quality Table

		Officer to be given a:		
	D10 Roll	**Corps**	**Divisional**	**Brigade**
Nationality		**command**	**command**	**command**
British, French	0-2	Poltroon	Poltroon	Poltroon
	3-4	Cautious	Poltroon	Poltroon
	5-6	Able	Cautious	Able
	7-8	Able	Able	Able
	9	Able	Gifted	Gifted
Russian, Prussian,	**0-2**	**Poltroon**	**Poltroon**	**Poltroon**
Spanish, Portuguese	3-4	Cautious	Cautious	Cautious
or other	5-6	Able	Able	Able
	7-8	Able	Gifted	Gifte
	9	Gifted	Gifted	Charismatic

EXAMPLE
The Prussian player rolls first for his commander-in-chief. As he is playing a corps-level game, he uses the corps command column and is fortunate to roll a 9 on the Commander Quality Table; his corps commander is gifted. Moving on to his two divisional commanders, he is less fortunate, rolling a 2 and a 4, so he will have to make do with one poltroon and one cautious commander. He can only hope his rolls in the brigade column will compensate...

These officers will have the following characteristics:

OFFICER TYPE	NUMBER OF LEADERSHIP FACTORS	DESCRIPTION	COMMAND RADIUS (inches)
Poltroon	2	An inexperienced officer, possibly a political appointee, not given to making sound decisions	8
Cautious	3	Nervous and lacking in confidence, slow to act, erring on the side of safety	8
Able	4	A steady and reliable commander. Capable of using initiative and giving clear direction	12
Gifted	6	Able and capable, often acts with flair and élan. Makes inspired decisions, good military knowledge	16
Charismatic	8	Inspires the loyalty and devotion of his subordinates, either through personality or proven skill	16

Leadership Factors

Each of the different characteristic types has a leadership factor. This relates to the number of units under command, initiative, lucky guesses and changes of orders. A commander can only command the number of units equal to his leadership factor. Therefore, any units not under command must test to see what they can do (see Out of Command below).

The use of leadership factors limits the control that a commander can have over a battle. If Murat is a charismatic commander, he can order up to eight subordinate formations under him. In other words, up to eight corps commanders. If we grade the Duke of Brunswick as an able corps commander, he can control up to 4 divisions as part of his command. A divisional commander graded a poltroon can command just two brigades with any effectiveness and a brigade level poltroon, just two regiments. The differing grades of commander come into operation with the larger battles, the more troops you have committed to battle, the more complex the command structure. For the most part, wargamers will be playing probably at corps or divisional level; so three or even two levels of command are usually more than sufficient.

Issuing Orders to Corps, Division and Brigade level

At the start of the game, the army commander of each side will issue the orders to their subordinates, so in a corps-level game he will be issuing them to divisional commanders, in a smaller game he will be a divisional commander issuing them to brigade commanders. These should be recorded in writing. They will take the form of allocating objectives, for example 'capture objective 2 and hold it', or 'advance on right flank supporting 2nd Brigade'.

Orders flow from the army commander through corps commanders and divisional commanders to the brigade commanders. The brigade commanders issue specific orders to each regiment or unit under their command. In every game there will be a series of objectives. These may be geographical locations on the battlefield. Sometimes, one side will not necessarily be performing any overtly aggressive movement; they may simply have to hold a position or, perhaps, conduct a fighting withdrawal. When orders are given to divisions or to brigades this places limitations and conditions on the brigade commander and what that brigade commander can tell regiments or units to do. The following table outlines the key overall orders that may be issued to divisions or brigades and the limitations that they place on the brigade commander:

GENERAL INSTRUCTION	LIMITATIONS ON THE RECEIVING OFFICER
Seize (specify objective)	If given to corps or division at least one subunit (division or brigade respectively) must be given seize orders, all others either support or manouevre orders. If given to brigade, at least one regiment in the brigade must be issued an assault order, all other regiments or units must be given either engage or advance orders.
Support (specify formation to support)	The formation is required to support another specified formation. If given to corps or division, no subunit may be given seize, hold or fighting withdrawal orders. If given to brigade, no regiments can be given assault, hold or retire orders.
Hold	The formation is expected to maintain control of an objective or position. If given to corps or division, at least one subunit must be given hold orders, none may be given fighting withdrawal or seize orders. If given to brigade, the brigade commander cannot issue any advance or assault orders and at least one regiment must be given a hold order.
Manoeuvre (specify general direction or route or eg turn left flank)	This order directs the formation to move along a particular route or in a particular direction. If given to corps or division, at least half its subunits must be given manoeuvre orders. If given to brigade, at least half units must be given advance orders.
Fighting withdrawal	The formation is instructed to slowly give ground to the enemy while avoiding being overwhelmed or outflanked. If given to corps or division, at least one subunit must have a fighting withdrawal order. If given to brigade, at least one unit must have a retire order.

Orders to corps, divisional and brigade commanders can be changed during the course of the game by means of couriers (see Changing Orders below)

Changing orders to corps, division and brigade commanders.

In principle, commanders are prevented from changing orders instantaneously based on guesswork or the wargamer's 'hand or eyes of God'. With the exception of the 'Lucky Guess' rule (see below) an army, corps or divisional commander can only change the orders given to a subunit once a request for such an order change has reached them from a subunit. It is easier for a brigade commander to change orders on his own initiative (depending on his quality), than it is for a divisional or higher commander. This is because he will be with the regiments of his brigade and therefore able to react to the changes in situation and circumstances, whereas a higher-level commander is more likely to be removed from the action and therefore less able to respond instantly to changes in circumstances or new opportunities or threats.

All of these commanders, including the brigade commanders, have a staff, which will include couriers. These couriers must transmit any information or requests for changes in orders. To request new orders a courier is dispatched to the next commander up in the chain of command. The courier model is placed on the table next to the commander and moved when it is that formation's turn to move (their card is drawn). When it delivers the message, the superior commander may, when it is his turn (ie his card is drawn from the deck), send him back with the new orders (expending one leadership factor in the process). No change can be made until the courier has returned to the subunit commander with the order change. It then has immediate effect. In the case of new orders to a brigade commander, he will not be able to act on them until the beginning of the following game turn when he issues new regimental orders. By contrast, a divisional commander receiving new orders from his corps commander, may immediately dispatch couriers with new orders to his brigades, unless his card has already been drawn this turn.

To reflect their position closer to the action, a brigade commander can attempt to act on their own initiative (allowing him to change the orders he received from his divisional commander). He must roll equal to or less than his leadership factor on a D10 (with '0' counting as zero). This is done at the beginning of the Game Turn, before issuing new regimental orders (see the Game Turn below). For example, a poltroon can change orders on a roll of 0,1 or 2 only. The reason why only brigade commanders are subject to this rule is that in normal circumstances the divisional commanders will have been ordered by higher command to carry out a specific set of orders, such as capture and secure a bridge or village and will not be expected to deviate from that order.

A brigade commander does not need to make such a roll to change the orders he gives to individual regiments (which he can do automatically at the beginning of each round, providing he has sufficient leadership points and they are within command radius, as discussed above); the test is to change the orders he received from above. So a brigade commander with orders to seize a hill must test like this to change his orders to 'Hold', which would allow him to issue different orders to his regiments.

Lucky Guesses

Only the overall army commander can make a lucky guess (which will be a corps or divisional commander depending on the size of the game).

Only one lucky guess can be made per turn; the player should use it when a situation occurs that has not been foreseen in the orders issued. It does NOT require a message to reach the army commander to trigger off a lucky guess opportunity. However, if the army commander does make use of a lucky guess then he must despatch a messenger to the relevant unit or sub-unit to pass the change of orders on to the officer in command of that unit. A lucky guess is made if the army commander rolls less than or equal to their leadership factor on a D10.

Brigade commanders issuing regimental orders

Once he has his orders from the divisional commander (or has successfully tested to act on his own initiative), the brigade commander then issues orders to his regiments, bearing in mind the restrictions outlined above. Regimental orders can be issued in the first phase of each game turn.

The appropriate order marker is placed face down next to each regiment or battery to conceal it from the opposing player until that unit take its turn in the action phase (feel free to photocopy the sheets provided at the back of this book).

Unless affected by a morale test or out of command result, regimental orders will remain in force in subsequent turns until the brigade commander changes them by issuing a new order.

Orders to regiments take the following forms.

Order	Purpose and limitations of the order
Assault	Advance aggressively towards the enemy, charge or attack, fight or shoot
Advance	Move forward towards the enemy, fight or shoot
Engage	Move as required, fight or shoot
Hold	Defend a position or objective, no movement, may fight or shoot
Retire	Move away from the enemy, fight or shoot if attacked

Note that only troops with assault orders may attempt to contact the enemy by charging.
Skirmishers cannot be given assault orders (and therefore cannot charge).

It takes one leadership factor to issue each order. Orders can only be issued to units within the commander's command radius (as shown on the commander quality table). Any units for which the brigade commander doesn't have sufficient points, or that are beyond his command radius, are 'out of command' (see Out of Command, below).

A brigade commander must attempt to change regimental orders to bring them into line with his orders from the division and give these changes priority in using up leadership factors. If the existing regimental orders of the out of command unit already comply with the brigade's orders, the brigade commander may simply leave them in force. Out of command, affects the ability to change a units orders, not of a unit to carry on with existing orders.

Out of Command

When a commander does not have sufficient leadership factors to give all of his units orders, or units are beyond his command radius, he may only change their orders by risking the Out of Command Table below. If their current regimental orders do not comply with the restrictions imposed by the brigade's orders, the brigade commander must attempt to change them.
If they receive a 'hold in place' result then they cannot initiate firing.

			D10 score		
Unit's training	**0-1**	**2-3**	**4-5**	**6-7**	**8-9**
Green	A	C	D	D	E
Trained	A	B	C	D	E
Experienced	A	C	E	F	F
Veteran	A	B	C	E	F
Elite	A	A	B	C	E

Out of Command table key

Result	Description
A	Obey orders as given.
B	Carry out previous orders or result (hold if first round).
C	Copy a neighbouring unit's actions if within 4 inches, otherwise hold.
D	Change to Advance orders, advance 4 inches this turn if possible.
E	Change to Retire orders, retire 4 inches this turn if possible .
F	Hold in place. Change to Hold orders and cannot initiate firing.

EXAMPLE

The Prussian commander von Sidnell is an able officer. A courier has arrived from his divisional commander, changing his orders from 'Hold' to'Seize objective A'. In the first phase of the next turn he must change the regimental orders to reflect the new orders from his superior - on seize orders at least one regiment must be given 'Assault' and the rest 'Advance' or 'Engage' orders. He has eight units under his command, but can only effectively command four of them. He uses up all of his leadership factors issuing orders to his cavalry and artillery. In order to change the orders of his four regiments of green cautious Landwehr, he must now risk the Out of Control table. He rolls a 1 for the 1st (obey orders) and he advances them 4 inches. He rolls a 3 for both the 2nd and 3rd Regiment; this means they can copy the 1st and advance 4 inches. For the 4th Regiment von Sidnell rolls an 8; the 4th decides to fall back 4 inches.

3. The Game Turn:

Game Turn Sequence

Follow the following stages in order for each game turn.

1. Brigade commanders may change orders.
2. Draw brigade card.
3. Challenge initiative (see challenge initiative)
4. Action phase. Player with initiative takes his turn.
5. Repeat steps 2, 3 and 4; continue to draw cards until all brigades have moved or taken action
6. Morale checks (see 8. Morale)
7. Compulsory moves
8. Shuffle deck ready for next turn.

3. 1 Change brigade orders, Issue regimental orders

If a courier delivered new orders from the divisional commander during the previous round, these now take effect. Alternatively, brigade commanders may attempt to change the orders they have from their divisional commander on their own initiative by rolling less than or equal to their leadership points on 1D10. If this fails the brigade's existing orders remain in force.

Brigade commanders may then change regimental orders to their units (regiments and batteries) in accordance with their existing orders. Each brigade commander first issues orders to those units within his command radius, prioritizing those whose orders are not in line with brigade orders, and expending one leadership factor for each until he runs out of these (see Command and Control).

To issue an order, the player simply places the relevant order marker (feel free to photocopy the sheet provided at the back of this book) next to the unit, face down to conceal it from the enemy until that unit takes action during the course of the game turn.

3.2 *Draw cards for initiative*
Draw the top card from the card deck to determine which units will move or fight next. If the opposing player does not wish to challenge for the initiative (3.3 below), then the player whose card was drawn may proceed to the action phase (3.4 below).

3.3 *Challenge initiative*
If the opposing player wishes to take the initiative – to forestall an enemy charge for example - he must make a leadership roll on a D10, rolling equal to or less than his leadership factor to succeed. The brigade commanders will generally do this. It can only be done once per turn. If the initiative challenge has been successful then the player who has won the initiative takes his turn. The relevant card is removed from the deck, while the first brigade card drawn (that of the player who lost the initiative) is returned to the deck, which is reshuffled.

For example, the card is drawn for a French brigade that contains units in position to declare a charge on a regiment of British infantry. The British commander has issued assault orders to this unit. He can seize the initiative back by making a successful leadership roll and then make his own charge. There would be little point in doing this if he had issued hold orders and he would have to await the charge resolve calculations to see how his troops respond to the attack. If he had issued retire orders, however, he could make the initiative challenge in the hope that he would be able to move his regiment out of danger of the charge. In winning an initiative challenge, the owner of the unit or units involved must act in accordance with the unit's existing orders. If the British challenge is successful, the relevant British card is removed from the deck, while the French card is returned to the deck and shuffled.

3.4 *Action Phase*
Once a formation has the initiative (its card was drawn and not successfully challenged), the controlling player may take action with it.

If the card drawn was a divisional or higher commander, then the controlling player may now move that commander; dispatch and move couriers with new orders to subunits or with requests for new orders to his superior; or fire a grand battery under his command.

If the active card was that of a brigade commander, the units under his command can take action according to their orders. Generally speaking, in a single action phase each unit may move and fire; or charge and fight; or continue to fight an ongoing melee.

The player with initiative should move all units that wish to move (declaring charges and testing resolve for attackers and defenders if they wish to move into contact), then conduct all firing, then fight all melees. See the sections on Movement, Charging, Firing and Melee.

Once in square, a unit cannot move and fire in the same turn, it can opt for either of these actions, but not both.

Artillery must be unlimbered to fire. It takes half a move to limber or unlimber. Therefore, a limbered battery moves 6 inches up a road, which is half its normal move. It then unlimbers. NB: It cannot fire until the next move. A battery may fire if it is already deployed. It can then limber but not move. Similarly, a stationary limbered artillery piece may unlimber and fire. A deployed battery may change its facing and fire or be manhandled forwards or backwards and fire.

3.5 - Repeat steps 3.2, 3.3 and 3.4

When the first brigade or other commander has had their turn, draw the next card from the deck and conduct the action phase for the next formation. When that formation has conducted their action phase, repeat, until all the cards have been drawn. When the last command drawn has performed its actions, move on to 3.6 Morale Checks.

3.6 - Morale Checks

Conduct any morale checks for casualties, seeing friendly units destroyed or routed, or attempts to rally from routing or recover from being shaken. See Section 8: Morale.

3.7 Compulsory Moves

Make moves resulting from morale checks, routing etc. Routers continue to move a full move away from enemy or towards own table edge until they are rallied or leave the field.

4. Movement

4.1 Movement Rates (inches)

	Road Track	Open	Rough Ford	Dense Wood	Wood	Swamp Marsh	Steep hill/ slope
Inf. Column	10	7	5	4	5	3	4
Infantry Line	4	4	3	2	3	2	2
Inf. Ext. Line	5	5	4	3	4	3	3
Inf. Square	3	3	2	•	•	•	•
Skirmishers	8	8	6	4	6	4	4
Cav. Column	20	12	8	▲	8	•	6
Cav. Line	8	8	6	▲	6	•	4
Dismounted Cav.	4	4	3	2	3	2	2
Art'y. Limbered	12	10	6	■	6	■	2
Art'y. Manhandled	3	2	2	■	1	■	1
Courier	20	20	20	20	20	20	20

▲ = Cavalry must dismount to pass through and move at dismounted cavalry rates
• = Impassable to cavalry or squares, must go round the obstacle
■ = Impassable to artillery, must go around the terrain feature

If a regiment marching along a road encounters a steep slope during that move, then add the 2 movement rates (i.e. road and steep slope) and divide by 2 to calculate the actual move. Other combinations are calculated in the same way, i.e. an open move into rough ground is the sum of 7 + 5 divided by 2 = 6 inches. The only exceptions are roads which pass through woodland and roads crossing rivers via bridges, in which cases there is no reduction to speed.

4.2 Wheeling and Oblique Movement

Regiments change direction by wheeling, this means pivoting the regiment on one flank whilst moving forward (or back) with the other flank. In effect, the regiment is moving on an arc. Wheels are measured from the outer edge of the arc, that is the corner moving furthest (the right corner of the line if wheeling left and vice versa). Experienced, Veteran and Elite regiments can wheel at full speed. Green and Trained regiments wheel at reduced speed, losing one quarter of their movement allowance. If a regiment is in extended line then no matter what its training grade, it will wheel at full speed.

Regiments that are in skirmish order do not have to wheel as they are in looser formations and can move in any direction the player chooses, so long as no individual base exceeds its movement distance.

Oblique movement involves changing the direction of march for a short distance but not the facing of the unit, effectively advancing diagonally. Units may oblique upto 45 degrees either side of straight ahead. Experienced, Veteran and Elite regiments must lose a quarter of their movement allowance to perform an oblique. Green and Trained regiments must lose half of their movement allowance to perform an oblique.

Anything not covered by wheeling and oblique movement constitutes a formation change, taking half a move to perform.

4.3 Formation changes

Infantry may change formation from column into line or line into extended line, and then into skirmish order. They may also change back from skirmish order to extended line, then into line and then from line into column. Infantry may form square from line or column, or from square to line or column. Infantry cannot change formation from column directly to extended line or vice versa. Except for forming square, the unit expends half of their current formation move in order to change formation. For example, a French regiment in extended line is entitled to move 12 cm. It wishes to change formation into line; therefore it can move up to 6 cm then change formation. Alternatively, it can change formation immediately and then move half of the move in the appropriate new formation.

Forming square is a special case, because it was usually only attempted under imminent threat of a cavalry charge when success or failure meant life or death for the regiment, so the exact timing could be of crucial importance for the outcome of the battle (and your game). The time taken to form square is dependent on the training of the unit, reflecting not only the complexity of the formation, but also the possibility of panic at the prospect of a cavalry charge.

Unit Training Grade	Time taken to form or come out of square
Green	2 turns
Trained	1.5 turns*
Experienced	1 turn
Veteran	Half turn*
Elite	Half Turn*

*This means that in the remaining half turn, the unit could move half a turn or fire.

Remember, for movement, defence and target purposes, if a square falls to a single stand, it is no longer considered to be a square. The owning player must immediately reform the unit into either a line or a column, facing whichever way he chooses, once this happens.

If forming different units into a single square, the stands of each unit must still be in base-to-base contact and must not be jumbled up in different faces of the square amongst stands of other units. The time taken would be determined by the unit with the lowest training grade.

Cavalry may also change formation from column to line or from line to column. The same procedure applies. When cavalry dismounts it must move up to half of its current mounted move stop then dismount without further movement. When dismounting, one base becomes horse holders. Cavalry may only remount if they are in base-to-base contact with their horse-holders. To remount a dismounted cavalry unit is assumed to be in base-to-base contact with their horse-holders. They can then mount and move half the distance of their column move.

Horse-holders move the same distance as dismounted cavalry. Dismounted cavalry must either move away from the enemy if they are within 4 inches of formed bodies of infantry or mounted cavalry, or they can remount and move away.

4.4 Moving through other units
Infantry in column or in line may move through an extended line of other infantry, losing 1 inch from their move. Infantry in line or extended line may also move through deployed artillery, again losing half an inch of their move. There are no other effects on either unit.

Any unit can move through troops in skirmish order with no penalty. If the main body passes through its skirmish screen, then the unit in skirmish order must halt, and use its turn to adopt the same formation as the main body.

4.5 Double Time
This applies only to limbered artillery or units in column on a road or in the open.

Double time allows units to be moved up to the front or redeployed to another flank during the battle. A brigade commander with regiments already in column may place a double marker next to each regiment in the order phase, before any movement takes place. This commits the regiment or regiments to make a double time move, during which they may be vulnerable to enemy action. Double time units may not move into combat or return fire.

Double time units must not come within 10 inches of enemy cavalry (unless a friendly unit is between them and the doubling unit) or 4 inches of an enemy unit in any circumstances.

Units must have one period of normal movement between each turn of double time. The double time unit rates (in inches) are:

Troop type	**Road**	**Open**	**Combination of road and open**
Infantry	14	10	12
Cavalry	28	20	24
Artillery	18	14	16

EXAMPLE

A brigade has just entered the table. The player wants to rush them up a road to get into action. In the Brigade Commanders Issue Orders phase, the player checks that there are no enemy units within the range limitations, then places a double marker beside the brigade. The unit will then move at double time when their brigade card is drawn.

5 Charges
Whenever a player wishes to move one unit into contact with an enemy unit, a charge is declared *instead* of making a normal move.

The procedure for declaring charges is:

1. Check regiment is within charge distance of intended target and eligible to charge.
2. Check resolve of charging unit and follow resultant instructions - take charge morale test? (see Testing Resolve).
3. Check resolve of defending unit and follow resultant instructions - take charge morale test? (Testing Resolve).
4. If appropriate, the charging regiment moves into base-to-base contact with the target regiment and they now engage in a melee.

5.1 Declaring a Charge

The charge distances below are instead of, not additional to, the normal move.

Charge Movement Table (inches)

	Road/Track/ Open	Rough	Woods	Downhill	Steep hill or slope
Infantry Line	6	5	5	6	3
Infantry extended	7	6	6	7	4
Infantry column	7	5	3	6	4
Cavalry Line	12	8	8	10	6
Cavalry column	14	7	7	14	7

If, upon checking, it turns out a unit is not within range, the unit may make a normal move instead.

If the unit is close enough to charge, check that it is not prevented by any of the following conditions.

Only regiments with assault orders may declare a charge.

Artillery, dismounted cavalry and skirmishers may never declare a charge and may not be given assault orders. Cavalry must remount or skirmishers reform into the appropriate formation before their unit can be given assault orders preparatory to charging.

Infantry and cavalry must be in line or extended line in order to declare a charge, with the exception of French infantry, who may charge in column.

Infantry will not declare a charge on mounted cavalry. During the Napoleonic Wars the practice was to attempt to deal with cavalry by firepower. Infantry, although less manoeuvrable than cavalry, presented a formidable challenge to cavalry and mounted troops were generally disinclined to charge home against formed infantry, but they may attempt it.

If charged whilst formed as skirmishers, a regiment will attempt to evade (fleeing 4 inches directly away from the charging unit).

Neither infantry nor cavalry may charge across a ford, swamp or marsh or bridge; nor may they initiate charges in dense woods.

Limbered artillery will always attempt to evade a charging unit. However, should they be contacted by a charging unit, they will be automatically captured.

5.2 *Testing Resolve*

The player controlling the charging unit rolls 1 D10 and consults the Attacker's Resolve Table (below) to get a letter code.

ATTACKER'S RESOLVE TABLE

Morale	Experience/Roll	0	1	2	3	4	5	6	7	8	9
IMPETUOUS	Green	A	A	A	A	A	K	K	M	N	N
	Trained	A	A	A	G	J	K	L	M	M	N
	Experienced	A	A	C	G	J	J	L	L	M	N
	Veteran	A	A	B	E	G	I	K	L	M	M
	Elite	A	A	B	B	E	G	H	K	L	M
CAUTIOUS	Green	A	A	G	K	L	L	M	M	N	N
	Trained	A	B	G	G	G	K	L	M	M	N
	Experienced	A	D	G	G	I	J	K	L	M	N
	Veteran	A	D	F	G	G	H	J	L	M	M
	Elite	A	B	D	E	E	G	H	K	L	M
STEADY	Green	A	A	A	G	K	K	M	M	N	N
	Trained	A	A	C	G	J	K	L	M	N	N
	Experienced	A	B	E	F	G	I	K	L	M	N
	Veteran	A	B	D	E	F	G	H	J	M	N
	Elite	A	B	B	B	B	D	G	H	I	M
CONFIDENT	Green	A	A	A	B	C	G	J	K	M	N
	Trained	A	A	B	C	F	G	H	L	M	N
	Experienced	A	B	B	E	F	F	G	K	M	N
	Veteran	A	B	B	D	D	E	F	H	I	L
	Elite	A	B	B	D	D	D	E	G	H	L
ARROGANT	Green	A	A	A	A	A	A	C	K	M	N
	Trained	A	A	A	A	C	C	K	M	M	N
	Experienced	A	A	A	B	B	E	H	L	M	M
	Veteran	A	A	A	B	B	B	E	H	L	M
	Elite	A	A	A	B	B	B	C	D	H	M

The letter code is then looked up on the Attacker's Resolve Table Key (below)

ATTACKER'S RESOLVE TABLE KEY

Result	Infantry	Cavalry
A	Charge home	Charge home
B	Attack at walk, no charge bonus	Charge home at trot - no charge bonus
C	Close to (or remain at) short range and fire. No charge	charge disordered - no charge bonus, penalty for disordered
D	Close to (or remain at) medium range and fire.	Charge disordered - no charge bonus, penalty for disordered
E	Close to (or remain at) long range and fire.	No charge. move 1/2 distance, become disordered
F	Morale test	move 1/2 distance, become disordered
G	Morale Test	Halt, do not charge
H	Pull back 1 inch and fire	Halt, do not charge
I	Halt at close range and fire - rout if counter charged	Break off charge and retire 2 inches
J	Halt and fire medium range - rout if counter charged	Break off charge - retire 4 inches
K	Halt and fire at long range - rout if counter-charged	Morale Test
L	Retire 2 inches facing the enemy	Morale Test
M	Retire disordered 2 inches, not facing the enemy	Rout
N	Rout	Rout

If the result indicates a morale test, this means a charge morale test must be taken *immediately*, using the Charge Morale Table (below) to obtain the final result of the attempt to charge.

The defender's resolve is now tested. Cross reference the defending unit's Morale and Training grades, and the appropriate column for its current orders. Use the same number rolled for the attacker's resolve table.

EXAMPLE

At the Battle of Raab, the Austrian 1st Line (Trained, Cautious) has assault orders and declares a charge on Greniers' 3rd Line (Experienced, Confident). Having checked that he is within charge distance (6 inches, since the regiment is in line on open ground), the Austrian player throws a D10 and refers to the Attacker's Resolve Table. The result is a 1, an A result. Referring to the key he is pleased to find this means the 1st Line will charge the French. The Austrian player now applies the same roll of 1 to the Defender's Resolve Table, using the appropriate column for the 3rd Line, who also have assault orders. The result is S1, which the key explains is Fire at Medium Range Effect then Fight. The French unit now fires at the charging Austrian unit before it is moved into contact (assuming it survives the volley) and a melee is fought.

DEFENDER'S RESOLVE TABLE

	ORDERS	ASSAULT				ADVANCE				ENGAGE				HOLD				RETIRE			
Morale	**Experience**	**0-3**	**4-5**	**6-8**	**9**	**0-3**	**4-5**	**6-8**	**9**	**0-3**	**4-5**	**6-8**	**9**	**0-3**	**4-5**	**6-8**	**9**	**0-3**	**4-5**	**6-8**	**9**
IMPETUOUS	Green	Q	R	Y2	Z	Q	R	T	Z	Q	W	Y1	Z	R	Y1	Y2	Z	R	Y2	Y3	Z
	Trained	Q	R	Y1	Y2	R	T	Y1	Y3	R	V	Y1	Z	S2	Y1	Y2	Z	S3	Y2	Y3	Z
	Experienced	Q	T	X	Y1	R	S3	W	Y2	S1	S2	V	Y3	S2	Y1	Y1	Z	S3	Y1	Y2	Z
	Veteran	R	S2	T	Y1	S1	S1	S3	Y1	S2	S2	U	Y2	S3	T	W	Y3	U	U	V	Z
	Elite	R	R	S1	T	S1	S2	S3	U	S1	S1	S2	V	S2	S3	T	X	S2	U	U	Y1
CAUTIOUS	Green	Q	S2	T	W	S2	T	V	W	S2	V	W	X	S3	T	W	Y3	W	Y1	Y2	Z
	Trained	R	S1	T	W	S1	V	W	W	S2	V	V	X	S2	T	V	Y2	T	W	Y2	Y3
	Experienced	R	S1	U	V	S1	S3	T	V	S1	T	U	W	S2	S3	T	X	S2	T	X	Y1
	Veteran	S1	T	U	U	S1	S2	T	V	S1	S2	T	V	S1	S2	S3	W	S2	S3	T	X
	Elite	S1	U	U	V	S1	S2	S3	V	S1	S2	S3	V	S1	S1	S2	W	S1	S2	S3	W
STEADY	Green	R	T	T	Y1	R	S2	T	Y2	S2	T	T	Y2	T	T	T	Y3	T	W	Y2	Z
	Trained	R	T	W	Y1	S2	S2	S3	Y1	S3	T	U	Y1	S3	S3	T	Y2	V	W	Y1	Y2
	Experienced	S1	S3	T	U	S1	S2	S2	U	S1	S3	T	V	S2	S2	S3	W	S3	V	W	Y1
	Veteran	S1	S2	S3	U	R	S1	S2	U	S1	S2	S3	U	S1	S1	S2	V	S2	T	U	W
	Elite	U	U	U	V	R	R	S1	V	S1	S1	S2	V	S1	S1	S1	V	S2	S3	T	V
CONFIDENT	Green	Q	R	S3	T	Q	R	T	W	R	S3	T	W	R	T	W	Y1	S2	S3	W	Y2
	Trained	R	S2	T	U	R	R	S2	U	R	S2	S3	U	S2	S3	T	V	S2	S3	V	W
	Experienced	S1	S2	S3	T	S1	S2	S3	T	S2	S2	S3	U	S2	S2	S3	V	S2	S3	T	V
	Veteran	S1	S2	U	U	S1	S1	S2	U	S1	S1	S2	U	S2	S2	S2	U	S2	S2	S3	V
	Elite	S1	S1	S2	U	S1	S1	S1	U	S1	S1	S1	U	S1	S1	S2	U	S2	S2	S2	U
ARROGANT	Green	Q	Q	R	T	Q	R	Y1	Y2	R	R	S3	W	R	S3	T	W	T	T	W	Y1
	Trained	Q	R	R	S3	Q	T	T	T	S2	S3	T	U	S3	S3	S3	V	S3	S3	V	W
	Experienced	R	S2	T	U	R	R	T	U	S1	S2	S3	U	S2	S2	S3	V	S2	S3	T	V
	Veteran	R	S1	U	U	R	R	S1	U	S1	S1	S2	U	S1	S2	S3	U	S2	S2	S2	V
	Elite	S1	S1	U	U	R	R	R	U	S1	S1	S1	U	S1	S1	S1	U	S1	S2	S3	V

DEFENDER'S RESOLVE TABLE KEY

Result	Infantry	Dismounted Cavalry	Mounted Cavalry	Artillery
Q	Counter-charge	Fire at full effect, then pull back half a move	Counter-charge	Fire double canister
R	Fire at short range effect, then fight	Pull back full move and remount	Counter-charge	Fire gun at short range effect
S1	Fire at medium range effect, then fight	Fire at half effect, then retire a full move	Receive charge stationary	Fire gun at medium effect
S2	Fire at long range effect, then fight	Retire full move disordered	Receive charge stationary	Fire gun at long range effect
S3	Stand and fight, fire if not charged	Morale Test	Receive charge stationary	Limber gun and move half a move
T	Morale Test	One stand flees, remainder fight if charged	Counter-charge disordered	Morale Test
U	Pull back half a move and fire	Two stands flee, remainder fight if charged	Morale Test	Defend gun, fight if charged
V	Pull back full move and fire	Retire full move facing the enemy, horses run off	Pull back half a move facing the enemy	Abandon gun, crew run away half a move
W	Retire not facing the enemy	Retire full move, not facing the enemy, horses run off	Pull back a full move facing the enemy	Spike gun and run away half a move
X	Retire disordered not facing the enemy	Unit is shaken, cannot move or fight	Retire full move not facing the enemy	Surrender if charged, spike and rout if not
Y1	One stand flees, remainder stands and fights	Rout	Retire full move disordered not facing the enemy	Spike gun and rout
Y2	Two stands flee, remainder stands and fights	Rout	One stand flees, remainder stands and fights	Rout leaving gun
Y3	Three stands flee, remainder stands and fights	Rout	Two stands flee, remainder stands and fights	Rout leaving gun
Z	Unit routs	Rout	Rout	Rout leaving gun

As with the attacker's resolve test, a 'morale test' result means a charge morale test must be taken immediately.

Charge Morale Test

If either the attacking (charging) or defending unit receives a 'morale test' result on their resolve table, they must immediately take a charge morale test. Roll 1 D10 and consult the Charge Morale Table to get a letter code.

CHARGE MORALE TABLE

		WITH ATTACKER				DEFENDER WITH ASSAULT ORDER				DEFENDER WITH ADVANCE ORDER				DEFENDER DEFENDER ENGAGE ORDER				WITH WITH HOLD ORDER				DEFENDER RETIRE ORDER			
MORALE	**EXPERIENCE**	**0-3**	**4-5**	**6-8**	**9**	**0-3**	**4-5**	**6-8**	**9**	**0-3**	**4-5**	**6-8**	**9**	**0-3**	**4-5**	**6-8**	**9**	**0-3**	**4-5**	**6-8**	**9**	**0-3**	**4-5**	**6-8**	**9**
IMPETUOUS	Green	A	A	G	J	B	C	E	F	B	C	E	F	B	C	E	J	C	E	J	L	E	F	I	L
	Trained	A	C	H	K	B	D	J	F	B	D	E	F	B	E	F	H	E	G	I	K	H	I	K	L
	Experienced	A	C	D	F	A	B	D	G	B	C	D	E	C	D	E	F	D	H	I	J	H	I	I	J
	Veteran	A	F	F	H	A	B	C	D	A	B	F	G	C	D	D	E	D	G	H	I	G	I	J	K
	Elite	A	B	C	I	A	C	D	G	A	B	D	F	A	B	F	G	D	F	G	H	G	H	I	K
CAUTIOUS	Green	A	B	E	H	B	C	G	K	C	E	J	K	E	G	H	I	E	I	J	K	H	I	K	L
	Trained	A	D	G	K	B	E	F	H	C	G	H	J	G	H	H	I	E	F	H	I	I	J	K	L
	Experienced	B	F	H	I	B	D	E	F	B	D	F	G	D	F	H	I	F	G	H	I	G	H	I	K
	Veteran	B	C	F	H	C	D	F	G	B	C	F	G	D	E	F	G	D	E	F	G	G	I	J	K
	Elite	B	F	F	I	A	C	D	F	A	D	E	G	B	C	D	F	C	D	E	F	F	G	I	K
STEADY	Green	A	E	G	L	B	C	D	G	B	E	F	G	B	D	I	K	E	I	J	K	H	I	K	L
	Trained	A	D	H	K	C	E	H	J	B	C	D	E	C	E	H	J	E	H	I	J	H	J	K	L
	Experienced	B	C	F	H	C	D	F	I	C	D	G	H	D	F	G	H	E	F	G	H	I	I	J	L
	Veteran	A	F	H	I	B	C	D	F	C	D	E	F	D	F	F	G	D	G	H	I	G	H	I	J
	Elite	A	F	H	I	A	D	F	H	C	D	F	G	D	E	F	F	D	E	F	G	G	G	H	I
CONFIDENT	Green	A	B	D	G	B	C	E	F	B	E	G	H	C	E	F	H	C	E	G	H	H	I	K	L
	Trained	A	C	G	H	C	D	F	G	B	D	G	I	C	D	F	G	D	G	H	I	E	I	J	K
	Experienced	B	C	F	I	A	C	D	F	B	C	D	F	B	C	D	F	D	E	G	H	E	H	I	J
	Veteran	A	C	F	F	A	B	D	F	A	C	D	E	B	C	D	E	D	E	G	I	E	G	I	J
	Elite	A	B	D	F	A	B	C	D	A	B	C	D	A	B	D	F	D	F	G	H	D	G	H	I
ARROGANT	Green	A	B	B	E	B	C	E	F	B	C	F	I	C	E	F	G	B	E	E	H	E	H	I	K
	Trained	A	B	E	F	B	D	F	G	B	D	F	H	C	D	E	F	D	E	G	H	H	I	J	K
	Experienced	A	A	C	F	A	B	D	G	A	C	D	E	B	C	D	E	C	D	E	F	G	H	I	J
	Veteran	A	B	F	F	A	A	C	D	A	B	D	E	A	D	E	F	C	D	F	G	D	G	H	I
	Elite	A	B	F	F	A	A	A	F	A	B	C	D	A	B	D	E	A	C	D	F	C	D	G	I

Look the letter code up on the Charge Morale Results Key:

Charge Morale Test Results Key

Result	Attacker Infantry	Attacker cavalry	Defender Infantry	Defender cavalry	Def. dismtd cav.	Defender Artillery
A	Charge to contact	Charge to contact	Counter-charge	Counter-charge	Fire short range	Fire double canister
B	Attack disordered	Attack disordered	Stand, fire if not charged	Counter-charge disordered	Fire long range, mount ,retire 4in	Ret by recoil 1 inch
C	Halt, fire short range	Attack disordered	Fire at short range, fight if charged	Receive at halt	Mount up and retire full move	Crew retreat 4in, abandon gun
D	Attack falters - test resolve again	Refuse to charge, halt	Fire at half effect using 2D10, fight	Receive at halt	Mount up and retire half move	Crew pinned, surrender if attacked
E	Fall back 4 inches	Fall back 6 inches	Hold orders changed to retire	About turn and retire for 8 inches	Retire half move on foot	Crew pinned - will be killed if attacked
F	Fall back 6 inches	Fall back 8 inches	Fall back 4 inches	Turn, retire 8in disordered	Retire full move on foot	Crew abandons gun and routs
G	One stand flees, remainder fire	Assault orders changed to advance	Fall back 6 inches	One stand flees-remainder fights	Horse's bolt, retire half a move on foot	Crew abandons gun and routs
H	Two stands flee, remainder fire	Turn about and retire a full move	One stand flees, remainder fights	Two stands flee, remainder fights	Horse's bolt, retire a full move on foot	Crew abandons gun and routs
I	Halt and fire at half effect using 2D10	Feint. Halt halfway to enemy	Two stands flee, remainder fights	Retire a full move	Panic and rout	Crew abandons gun and routs
J	Assault orders changed to Advance	Feint. Halt halfway to enemy, disordered	Three stands flee, remainder fights	Evade full move and scatter - 2 turns to reform	Panic and rout	Crew abandons gun and routs
K	Panic and rout	Panic and halt shaken	Panic and rout	Panic and rout	Panic and rout	Crew abandons gun and routs
L	Panic and rout	Panic and rout	Panic and rout	Panic and rout	Panic and rout	Crew abandons gun and routs

EXAMPLE

The elite, arrogant 7th French Hussars need to take a charge morale test as indicated in their Attacker's Resolve result. They are facing an Austrian trained, steady 2nd Hussars regiment. Both units have assault orders. The French player throws a 2. The result is A – charge to contact. If the defending unit also had to take the morale test, then the Austrian player would also have thrown a D10. In this case, a 4 is thrown. This is an E result – which means they must attempt to evade the French by turning about and retiring 8 inches.

6. Firing

Who can fire?

To fire, the target must be in range of the firer, within their arc of fire and, obviously, in line of sight. The arc of fire is forty-five degrees either side of straight ahead for all units, including artillery. Arc and range can be measured from the nearest front corner of the firing unit to the nearest part of the base of the target unit.

Line of sight is blocked by hills, buildings, woods (unless the enemy are in firing positions on the edge of the wood) and by all friendly units. Line of sight is also blocked by enemy units other than the target, except in the case of skirmishers who are too close (less than 2 inches) to a formed regiment behind them. In this case, the unit behind may be targeted.

Although firing is resolved on a unit basis, line of sight is determined for each base individually.

For example, a five-stand regiment might have three stands that could fire, and two that could not due to blocked line of sight. This is taken into account in the firing procedure below.

Instances of mounted cavalry firing were rare in the Napoleonic period. For simplicity, only dismounted cavalry may fire.

Firing Procedure.

1. Check the range to the target (short, medium or long?)
2. Determine number of firing dice (minimum of 1)
3. Roll dice and refer to Musketry Fire Table or Artillery Fire Table as appropriate
4. Check result against Firing Results Key.
5. Conduct any morale checks specified on Firing Results Key.

6.1 Checking the range

Check the range in inches. Measure from front of base to nearest point of target unit's base. Using the chart below, determine whether it is short, medium or long range.

Weapon Ranges (in inches)

Weapon	Short Range	Medium Range	Long Range
Pistol	0 - 1	over 1 - 2	over 2 - 3
Cavalry Carbine	0 - 3	over 3- 5	over 5 - 7
Musket	0 - 4	over 4 - 6	over 6 - 8
Rifle	0 - 5	over 5 - 8	over 8 - 12
Light Artillery Shot	0 - 8	over 8 - 16	over 16 - 24
Light Artillery Shell	0 - 8	over 8 - 16	over 16 - 24
Light Artillery Canister	0 - 4	-	-
Medium Artillery Shot	0 - 12	over 12 - 20	over 20 - 32
Medium Artillery Shell	0 - 12	over 12 - 20	over 20 - 32
Medium Artillery Canister	0 - 6	-	-
Heavy Artillery Shot	0 - 16	over 16 - 28	over 28 - 40
Heavy Artillery Shell	0 - 16	over 16 - 28	over 28 - 40
Heavy Artillery Canister	0 - 8	-	-

6.2 Determine number of firing dice

For infantry and dismounted cavalry fire, use the following table. Each different experience type is allocated a basic number of D10s to use each time it fires, the number of D10 thrown can be amended by certain factors or modifiers (see Firing Modifiers).

Infantry or dismounted cavalry Type	Number of D10
Green	1
Trained	2
Experienced	3
Veteran	4
Elite	5

For artillery, the basic number of firing dice before modifiers is determined by calibre and ordnance being fired. For simplicity, cannon fire canister or ball, howitzers fire shell.

Artillery Type and Ordnance Being Fired	Number of D10
Light shot	1
Light shell	2
Light canister	2
Medium shot	2
Medium shell	3
Medium canister	3
Heavy shot	3
Heavy shell	4
Heavy canister	4

Once you have the basic number of firing dice, apply the following modifiers.

Firing Modifier	Conditions	Change to D10 rolled
Aimed Fire against stationary target	To qualify the firing unit must have been stationary itself in this and the previous turn. Artillery must not have changed facing.	+2D10 per unit/gun
Rifle or Carbine Fire	To qualify, the firing unit must not have moved this turn	+1D10 per stand
Target is in cover	Target must be in solid protection such as walls earthworks/buildings or dense woods	-1D10 per unit/gun
Target is moving	Target must have moved more than a half move this turn. Artillery changing facing does NOT count.	-1D10 per unit/gun
Firer is disordered	Firer must be currently disordered	-1D10 per unit/gun
Per Stand unable to fire	Stands previously removed or with no line of sight.	-1D10
Firer has moved	Firer must have moved more than a half move this turn. Artillery has changed facing or moved at all.	-1D10 per unit/gun

Regardless of modifiers, the firing unit will always get a minimum of one firing die.

Targetting enemy commanders - If any enemy commander is within 2 inches of the target unit, the firer may allocate one die to attempting to hit him. See Section 9: Risk to Commanders.

6.3 Roll Dice and Refer to Fire Table

The firing player now rolls the modified number of dice. He must select the result of just ONE of those dice (obviously the best one, usually the lowest) and refers to the Musketry Fire Table or the Artillery Fire Table as appropriate.

The firer's experience and the die roll give the row, and the target's formation and range give the column. This will give you a letter code which can then be checked on the Firing Results Key.

On the firing tables the range is indicated by s = short range, m = medium range, l = long range and can = canister (artillery only).

Musketry Fire Table

		Target type and formation																					
		Infantry Line			Infantry Column or Square			Infantry Extended Order			Dismounted Cavalry or Skirmishers			Mounted Cavalry			Deployed Artillery			Limbered Artillery			
Firing Unit's Experience	**D10 roll**	**S**	**M**	**L**	**S**	**M**	**L**	**S**	**M**	**L**	**S**	**M**	**L**	**S**	**M**	**L**	**S**	**M**	**L**	**S**	**M**	**L**	**D10 roll**
Green 1D10	0-2	A	B	B	A	A	B	B	B	C	B	B	D	A	A	B	A	B	C	A	C	D	0-2
	3-4	B	B	C	A	B	C	C	D	E	B	C	F	A	B	C	B	C	D	B	D	E	3-4
	5-6	C	D	D	C	C	E	D	E	F	C	D	F	B	C	D	C	D	E	C	E	F	5-6
	7-8	D	E	F	D	E	F	E	E	F	D	E	F	C	D	E	E	F	F	D	F	F	7-8
	9	E	F	F	E	F	F	F	F	F	E	F	F	D	E	F	F	F	F	E	F	F	9
Trained 2D10	0-2	A	A	B	A	A	B	A	B	C	A	B	C	A	A	B	A	A	B	A	B	C	0-2
	3-4	B	B	C	A	B	C	B	C	D	B	C	D	A	B	C	A	B	C	B	C	D	3-4
	5-6	B	C	D	B	C	D	C	D	E	C	D	E	B	C	E	B	C	D	C	D	E	5-6
	7-8	C	D	E	C	D	E	D	E	F	D	E	F	C	D	E	C	D	E	D	E	F	7-8
	9	D	E	F	D	E	F	E	F	F	E	F	F	D	E	F	D	E	F	E	F	F	9
Experienced 3D10	0-2	A	A	B	A	A	B	A	A	B	A	B	C	A	A	A	A	A	A	A	B	B	0-2
	3-4	A	B	C	A	B	B	B	B	C	A	C	D	A	B	B	A	B	B	A	B	C	3-4
	5-6	B	B	D	A	B	C	C	D	E	B	D	E	B	C	C	B	C	C	B	C	D	5-6
	7-8	C	C	E	B	C	E	D	E	F	C	E	F	C	D	E	C	D	E	C	D	E	7-8
	9	D	E	F	C	D	F	E	F	F	D	F	F	D	E	F	D	E	F	D	E	F	9
Veteran 4D10	0-2	A	A	B	A	A	A	A	A	B	A	A	B	A	A	A	A	A	A	A	A	B	0-2
	3-4	A	A	B	A	A	B	A	B	C	A	B	C	A	A	B	A	A	B	A	B	C	3-4
	5-6	B	B	C	A	B	C	B	C	D	B	C	D	B	B	C	A	B	C	B	C	D	5-6
	7-8	B	C	E	B	B	E	C	D	E	C	E	F	B	C	D	B	C	D	C	D	E	7-8
	9	C	D	F	C	C	F	D	E	F	D	F	F	C	D	E	C	D	E	D	E	F	9
Elite 5D10	0-2	A	A	A	A	A	A	A	A	A	A	A	A	A	A	A	A	A	A	A	A	A	0-2
	3-4	A	A	B	A	A	B	A	B	B	A	B	B	A	A	B	A	B	B	A	B	B	3-4
	5-6	A	B	B	A	B	C	B	C	C	B	C	D	B	B	C	A	B	C	B	C	C	5-6
	7-8	B	C	E	A	C	D	B	D	E	D	E	F	C	C	E	B	C	D	C	E	E	7-8
	9	B	D	F	B	D	E	C	E	F	E	E	F	D	E	F	C	D	E	D	F	F	9

Artillery Fire Table

			Target type and formation							
Artillery Unit's Experience	**D10 roll**	**CAN**	**Infantry Line S ML CAN**	**Infantry Column or Square S ML CAN**	**Infantry Extended Order S ML CAN**	**Dismounted Cavalry or Skirmishers S ML CAN**	**Mounted Cavalry S ML CAN**	**Deployed Artillery S ML CAN**	**Limbered Artillery S ML**	**D10 roll**
Green	0-2	B	A B C A	A B B C	C D E C	B C D B	A B C A	A C D A	A B C	0-2
	3-4	C	B C D B	A C C D	D E F D	C D E C	B D E B	B D E B	B C D	3-4
	5-6	D	C D E C	B D E E	E E F E	D E F D	C E F C	C E F C	C D E	5-6
	7-8	E	D E F D	C E F E	F F F F	E F F E	D F F D	D F F D	D E F	7-8
	9	F	E F F E	D F F F	F F F F	F F F F	E F F E	E F F E	E F F	9
Trained	0-2	B	A B C A	A A B B	A B C B	A B C A	A B C A	A B C A	A B C	0-2
	3-4	B	B B D B	A A C C	B C D C	B C D B	A C D A	A C D B	B C D	3-4
	5-6	C	B C E B	A B D D	C D E D	C D E C	B C D B	B D E C	C D E	5-6
	7-8	D	C D F C	B C E E	D E F E	D E F D	C D E C	C E F D	D E F	7-8
	9	E	D E F D	C D F F	E F F F	E F F E	D E F D	D F F E	E E F	9
Experienced	0-2	A	A B B A	A A A A	A A B A	A A B A	A A A A	A A A A	A B B	0-2
	3-4	B	A B C A	A A B B	B B C B	B B C B	A B B A	A B C B	B C D	3-4
	5-6	C	B D D B	A B C C	C D D C	C C D B	B C D B	B C D C	C D E	5-6
	7-8	D	C E F C	B C D D	D E F D	D E F C	C D E C	C D E D	D E F	7-8
	9	E	D F F D	C D E E	E F F E	E F F D	D E F D	D E F E	E F F	9
Veteran	0-2	A	A A B A	A A A A	A A A A	A A A A	A A A A	A A A A	A A B	0-2
	3-4	B	A B C A	A A B A	A B B A	A B B A	A A B A	A B B A	A B C	3-4
	5-6	B	B C D B	A B B B	B C C B	B C C B	A B C B	B C E B	B C D	5-6
	7-8	C	B D E B	A B C C	C D E C	C D E C	B C D C	C D E C	C D E	7-8
	9	D	C E F C	B C D D	D E F D	D E F D	C D E D	D E F D	D E F	9
Elite	0-2	A	A A B A	A A A A	A A A A	A A A A	A A A A	A A A A	A A A	0-2
	3-4	A	A A C A	A A B A	A A B A	A A B A	A A A A	A A B A	A A B	3-4
	5-6	B	A B D A	A B B B	B B C B	B B C A	A B B A	A B C A	A B C	5-6
	7-8	C	B C E B	A B C B	C C D B	C C D B	B C C B	B C D B	B C D	7-8
	9	D	B D F C	B B D C	D D E C	D D E C	C D E C	C D E C	C D E	9

6.4 Check Results against Firing Results Key

Once you have obtained a result code from the relevant firing table, check what that result means on the Firing Results Key.

Firing Results Key

Result	**Musketry Fire Infantry or Cavalry Target**	**Artillery Target**	**Artillery Fire Canister against Infantry or Cavalry**	**Shot or Shell against Infantry or Cavalry**	**Any fire against artillery**
A	Two stands killed, morale test	Three crew killed, morale test	Three stands killed, morale test	Two stands killed, morale test	Gun is destroyed, one crew killed, morale test
B	One stand killed, no morale test	Two crew killed	Two stands killed	One stand killed and morale test	Two crew killed
C	Target unit is shaken it must halt. No combat for one turn	One crew killed	One stand killed	Target shaken, no firing or movement for one turn	One crew killed
D	Target unit is disordered	Target shaken, no firing or movement for one turn	Target shaken, no firing or movement for one turn	Target unit is disordered	Target shaken, no firing or movement for one turn
E	Target unit takes a morale test	Crew takes a morale test	Target unit is disordered	Target unit takes a morale test	Target unit takes a morale test
F	No effect	No effect	Target unit takes a morale test	No effect	No effect

If the result indicates a morale test, the target unit must take this immediately.

Always take a morale test if you lose two or more casualties in a single phase (eg two units fired at you, each scoring a result of '1 stand killed, no morale test').

EXAMPLE OF FIRING

The Austrian 1st Grenadiers, a veteran steady unit, is firing at an advancing French infantry regiment in column. The distance is 5 inches, so this is medium range. As veterans, the grenadiers start with 4 D10. The target is moving and the grenadiers have also moved, so they lose 2 D10, leaving them with 2 D10. The Austrian player throws a 6 and an 8. He opts to use the 6. Using the Musketry Fire Table, we cross reference veteran with infantry column, medium range and the roll of 8. This gives us a B result which is one stand killed.

Targetting enemy commanders - Note that if an enemy commander is within 5cm of the target unit he may be at risk of being hit. See Risk to Commanders, below.

7: Melee or Close Combat

Once units are in base-to-base contact, close combat can take place. Melees in these rules are designed to be short and bloody. In most cases, a definite result will be obtained, although in rare cases the combat may continue into the next move.

Melee sequence

1. Determine number of dice attacker rolls, then roll them.
2. Select result(s) to apply - usually single best (lowest) result, but modifiers may allow more. Refer to Attacker's Melee Resolution Table and Melee Results Key.
3. Before removing any casualties, repeat the process for the defender, using the Defender Melee Resolution Table and applying modifiers as appropriate.
4. Conclude Combat: remove casualties from both sides, move units pushed back etc
5. Conduct melee morale test.

7.1 Determine number of attacker's dice

The basic number of dice a unit will roll is determined by their training and experience: Green, 1D10, Trained 2D10, Experienced 3D10, Veteran or Elite 4D10. This number may be adjusted by certain modifiers (see the Melee Modifiers Table below), but the attacker will always throw a minimum of 1 die.

As far as melees are concerned, artillery batteries fight as if they were Trained infantry; one stand of guns is equal to one stand of Trained infantry.

Dismounted cavalry always fight as Trained infantry in melees, regardless of their actual training status.

7.2 Check and select results to apply.

As with firing, the numbers on the dice are not added up, and the attacker must choose the single best (lowest) result obtained by his throws. Some melee modifiers allow the defender to discard some of the attacker's rolls, while others allow the attacker to apply more than one result.

Melee Modifiers Table

Attacker modifiers	**Defender modifiers**
Charging count results of 2 D10 rolls	Counter-charging cancels out the attackers extra D10 result for charging
Attacking flank count the results of 3 D10 rolls	If behind defences the defender can select and reject the results of two of the attackers D10 rolls
For each stand less than full strength, defender can opt to disregard that number of attackers D10 rolls	Disordered reduce the number of rolls on Defender's Melee Resolution Table by one die.
Disordered reduce the number of rolls by one	Infantry in square against cavalry may have an extra D10 or cancel the charge bonus of the attacking cavalry
Cavalry facing an infantry square of 6 stands or more lose one roll	Extended line versus line or column, lose 1D10

Example: A veteran infantry regiment attacks an infantry regiment defending a redoubt. As veterans they get to roll 4 dice, scoring 1, 4, 7 and 9. The defender's are behind defences so they can reject two of these, leaving the attacker with the 7 and 9, but as the attacker's charged this turn they may count both results.

The selected dice roll(s) are then applied to the Attacker Melee Resolution Table, cross-referencing the attacker's experience and the appropriate column for the type of attacking and defending troops (eg infantry v cavalry). This will give a code letter which can be looked up on the Attacker column of the Melee Results Key to find the actual effect.

Attacker Melee Resolution Table

Attacker Infantry or Cavalry	D10 roll	Infantry v Infantry	Infantry v Cavalry	Cavalry v Infantry	Cavalry v Cavalry	Infantry v Artillery	Cavalry v Artillery
Green 1D10	0-2	B	D	B	B	A	A
	3-4	D	E	C	D	B	A
	5-6	E	G	D	E	C	B
	7-8	F	H	G	F	E	C
	9	H	I	J	J	E	D
Trained 2D10	0-2	A	B	B	B	A	A
	3-4	C	D	C	C	B	A
	5-6	D	E	D	E	C	B
	7-8	E	G	D	F	D	B
	9	F	H	E	G	E	C
Experienced 3D10	0-2	A	B	A	B	A	A
	3-4	B	C	B	C	A	A
	5-6	C	D	C	D	B	B
	7-8	D	E	D	E	C	C
	9	E	F	E	F	D	D
Veteran 4D10	0-2	A	B	A	A	A	A
	3-4	A	C	A	B	A	A
	5-6	B	D	B	C	B	A
	7-8	D	D	C	D	B	D
	9	E	E	D	E	C	C
Elite 4D10	0-2	A	B	A	A	A	A
	3-4	A	C	A	B	A	A
	5-6	B	D	B	C	A	A
	7-8	B	D	B	C	A	A
	9	C	E	C	E	C	C

Melee Results Key

Result	Attacker	Defender	Result
A	Two defending stands killed	Two attacking stands killed	A
B	One defending stand killed	One attacking stand killed	B
C	One defending stand flees	One attacking stand flees	C
D	Defender pushed back 4 inches	Attacker pushed back 4 inches	D
E	Defender takes morale test	Attacker takes morale test	E
F	Stalemate	Stalemate	F
G	Attacker pushed back 4 inches	Defender pushed back 4 inches	G
H	One attacking stand killed	One defending stand killed	H
I	One attacking stand flees	One defending stand flees	I
J	Attacker takes morale test	Defender takes morale test	J

Example: Continuing the previous example, the owner of the veteran infantry unit finds the column for infantry vs infantry and finds that his remaining scores of 7 and 9 give a D and an E result, respectively. The Attacker column of the Morale Results Key explains that this means the defenders are pushed back 4 inches and must also take a morale test.

If an enemy commander is with the target unit, the attacker may exchange one stand killed for a chance of killing the commander, see Section 9: Risk to Commanders.

7.3 Repeat Process for the Defender

Once the attacker's rolls have been resolved, we must determine how effective the defender's efforts have been. This is done before any casualties caused by the attacker are removed as the two sides would, of course, be fighting simultaneously. Repeat steps 1-3 but using the Defender Melee Resolution Table and the Defender column on the melee results key.

Defender Melee Resolution Table

Defender Infantry or Cavalry	D10 roll	Infantry v Infantry	Infantry v Cavalry	Cavalry v Infantry	Cavalry v Cavalry
Green 1D10	0-2	B	D	B	A
	3-4	C	E	C	B
	5-6	D	G	E	D
	7-8	E	H	G	E
	9	F	I	I	F
Trained 2D10	0-2	A	B	A	A
	3-4	C	D	B	B
	5-6	E	E	D	C
	7-8	F	G	E	E
	9	F	H	G	F
Experienced 3D10	0-2	A	B	A	A
	3-4	B	C	B	B
	5-6	E	E	D	C
	7-8	F	G	E	D
	9	F	H	G	B
Veteran 4D10	0-2	A	B	A	A
	3-4	B	C	B	B
	5-6	C	D	D	B
	7-8	D	E	E	C
	9	E	G	H	D
Elite 4D10	0-2	A	B	A	A
	3-4	B	C	B	B
	5-6	C	E	C	C
	7-8	E	G	E	E
	9	G	H	H	F

EXAMPLE

The Austrian 2nd Line (Trained, Steady) at full strength charges the French 1st Grenadiers (Experienced, Arrogant). The French counter-charge. There are no modifiers to the D10 rolls. The Austrians throw a 7. We refer to the Attackers Melee Resolution Table, cross referencing Trained, a roll of 7 and use the infantry v infantry column. This gives a result of an E in the attacker column which means a compulsory morale test of the defender of the melee (the French). The French player then rolls a 5 and uses the Defender Melee Resolution Table, he cross references experienced, a roll of 5 and also uses the infantry v infantry column. The result is also an E in the defender column which means that the attacker must also take a morale test.

7.4 Conclude Combat.

Now remove all casualties and move units pushed back, fleeing etc.

Following the calculation of the results, it will be determined if there is a winner or a loser of the melee. This should often be very clear from the relative casualties inflicted (which are now removed) or as a result of one of the regiments being forced back.

Melee only lasts for one turn, except in the rare cases that both sides produce just a Stalemate result. At the end of the melee, whether a stalemate result has been produced or not, both sides will test their morale, the only exception to this is when one side receives a Morale Test result from the Melee Resolution Table and other does not.

When a stalemate result is produced, then both sides will break off and pull back 2 inches. A unit that was defending an obstacle can reoccupy their position in their next turn provided they have been given an Assault, Advance or Engage order. If they have a Hold order, they will remain in place and not reoccupy the defensive position until their orders have been changed.

Both sides after a melee are also considered to be disordered and must reform.

8 Morale

There are five different morale grades; each unit is assigned a morale grade before the game begins. In the scenarios provided, we have allocated morale grades to regiments on the basis of their historical performance in battle. Morale is distinct from a unit's training as it represents their level of reliability, regardless of the training they may have received. The five morale grades are described in the following table:

Morale Grade	Basic Factor	Artillery	Infantry	Cavalry	Description of morale type
Impetuous	9	Not allowed	Allowed	Allowed	Rash and volatile, their morale can be quite fragile and unpredictable
Cautious	8	Allowed	Allowed	Allowed	Brittle and often nervous, a degree of inexperience often unwilling to take chances
Steady	10	Allowed	Allowed	Allowed	Reliable and capable troops
Confident	11	Allowed	Allowed	Allowed	Will generally follow orders, very competent troops
Arrogant	12	Allowed	Allowed	Allowed	Over confident, proud, lacking in caution and disdainful of their opponent's abilities

When to test morale

Firing, charging and melee results will indicate when a regiment or unit has to test its morale, and those tests are taken straight away. However, there are other circumstances when morale tests should be taken at the end of the game turn:

- If a regiment or unit from its own brigade or division has routed and passed with 12 inches of the unit.
- If the unit is trying to recover from being shaken, disordered or having been routed.
- If the unit has seen a friendly unit destroyed by the enemy within 12 inches
- If the unit has seen its brigade, division or army commander killed or wounded within 12 inches.
- In a game turn in which a regiment or unit has taken casualties and has not be required to take a morale test as a result of any other circumstances that turn.

Other optional morale tests can include (if players mutually agree before game commences)

- Green or Trained units under fire (regardless of any casualties inflicted) for the first time.
- Impetuous units that see an enemy unit retreat, retire or rout within 12 inches.
- Arrogant units that have inflicted casualties on an enemy unit in this game turn.

Morale tests resulting from the charging procedure (see 5.2 Testing Resolve) are taken using the Charge Morale Table (p.??). All other morale tests are taken as follows.

How to test morale

When required to test a unit's morale, the player rolls 2D10. To pass the morale test, the player must score equal to or less than the Basic Factor (BF). For this purpose, a 0 on a D10 equals 10, in other words rolling a 0 and 6 equals 16, and two 0's equals 20. Modifiers are added or subtracted from the total as appropriate. If the final score is higher than the Basic Factor, the unit has failed its test, but the player must calculate by how much it is higher. It is this difference that gives the result on the morale table.

For example, if the testing unit is Steady, its BF is 10, the player rolls an 8 and a 7 which equal 15, and therefore he has exceeded the unit's BF by 5. The player would then check the current orders of the unit, which in this case is Hold and the experience of the unit, which in this example is Trained. Referring to the Morale Table, the player then finds the Steady – Trained line and cross-references this with the Hold column and finds the 5-8 columns. The result is E which states that the player must replace the unit's current Hold orders with a Retire order. In many cases there may be other factors that need to be taken into account, these are applied to the player's 2D10 roll, either added to the score or subtracted from the score.

Morale Modifier Table

Add 1 to the 2D10 roll	**Subtract 1 to the 2D10 roll**
For each stand lost by the unit to date	If unit is at full strength
If the brigade commander has been killed this turn	If the brigade commander is within 5 inches
If unit are Raw Trained or Raw Green troops under fire for the first time	If there are friendly troops within 4 inches
If there are no friendly units with 4 inches	If unit is in cover
If fired on or attacked from the flank or rear	If there are routing or retiring enemy units within 4 inches
If currently disordered or shaken	If the unit has been victorious in melee or combat
If an enemy unit is within 4 inches	If the divisional commander is within 5 inches
If the unit has been pushed back in a melee or combat	If the army or corps commander is within 6 inches
For each friendly unit seen destroyed, routing or retiring within 5 inches	If the unit has inflicted casualties on the enemy this turn

EXAMPLE

A British impetuous cavalry regiment needs to take a morale test. Its basic factor (BF) is 9. Relevant modifiers are 'no friendly units within 4 inches' and ' unit is currently shaken'. Both of these factors add to the 2D10 roll. The British player rolls and 6 and 7, then adds 2 giving a total of 15. He has exceeded his BF by 6. The regiment has Engage orders and are veterans. We cross reference these factors on the Morale Table to reveal an F result. For cavalry, this means that they will have to retreat 2 inches not facing the enemy.

Morale Table

		Assault Order					Advance Order					Engage Order					Hold Order					Retire Order				
Morale	**Experience**	**1-4**	**5-8**	**9-12**	**13-16**	**17-20**	**1-4**	**5-8**	**9-12**	**13-16**	**17-20**	**1-4**	**5-8**	**9-12**	**13-16**	**17-20**	**1-4**	**5-8**	**9-12**	**13-16**	**17-20**	**1-4**	**5-8**	**9-12**	**13-16**	**17-20**
Impetuous	Green	A	C	F	G	H	A	A	B	D	F	A	A	B	F	G	A	B	B	E	G	A	B	F	G	H
BF9	Trained	A	B	C	F	H	A	B	D	F	G	A	B	F	G	H	B	E	F	G	H	B	F	G	H	I
	Experienced	B	B	G	H	I	B	D	F	G	H	B	F	G	H	I	E	F	G	H	I	F	G	H	I	J
	Veteran	B	C	H	H	I	D	F	G	H	I	B	F	F	G	H	F	G	G	H	I	G	G	H	I	J
	Elite	B	C	C	H	I	F	G	H	I	J	B	F	G	G	H	G	H	H	I	J	H	H	I	J	J
Cautious	Green	B	C	F	H	I	B	B	D	F	G	A	B	F	G	H	B	E	G	H	I	F	G	H	I	J
BF8	Trained	B	F	G	H	I	B	D	F	G	H	B	F	G	H	I	E	G	H	H	I	G	H	I	I	J
	Experienced	A	B	C	H	I	D	F	G	H	I	F	G	H	I	I	F	G	H	I	J	H	H	I	J	J
	Veteran	A	C	G	H	I	D	G	H	I	J	F	H	I	I	J	G	H	I	I	J	H	I	I	J	J
	Elite	B	C	G	H	I	G	H	I	I	J	H	H	I	J	J	E	G	H	I	J	I	I	J	J	J
Steady	Green	A	A	B	F	G	A	B	D	G	H	B	F	G	H	I	A	E	F	G	H	B	G	H	I	J
BF10	Trained	A	B	C	G	H	B	D	F	G	H	F	G	H	I	J	B	E	E	F	G	F	G	H	I	J
	Experienced	B	C	F	H	I	B	F	G	H	I	F	F	G	H	I	E	F	G	H	I	G	H	H	I	J
	Veteran	B	C	G	H	I	A	B	F	G	H	A	B	F	G	H	E	G	H	I	J	H	H	I	I	J
	Elite	A	B	F	G	H	B	F	G	H	H	B	F	G	H	I	E	G	H	I	I	H	I	I	J	J
Confident	Green	A	A	B	F	G	A	B	D	F	G	A	A	B	B	F	A	B	B	E	F	B	B	F	G	H
BF11	Trained	A	A	C	G	H	B	D	F	G	H	A	B	F	F	G	A	B	E	F	G	B	F	G	H	I
	Experienced	A	B	B	C	F	B	F	G	H	I	A	B	F	G	H	B	E	F	G	H	F	G	H	I	J
	Veteran	A	B	C	F	G	B	D	F	G	H	B	B	F	G	H	E	F	G	H	I	G	G	H	I	J
	Elite	A	C	F	G	H	D	F	G	H	H	B	F	G	H	I	F	G	H	I	J	G	H	I	I	J
Arrogant	Green	A	A	A	B	C	A	A	A	B	D	A	A	B	B	F	A	B	E	F	G	A	B	F	G	H
BF12	Trained	A	B	B	C	F	A	A	B	D	F	A	B	B	F	G	A	E	F	G	H	A	F	G	H	I
	Experienced	A	B	C	F	G	A	B	D	F	G	A	B	F	G	H	A	B	E	F	G	A	G	H	I	J
	Veteran	A	C	F	G	H	B	D	F	G	H	A	F	G	H	I	A	B	B	E	F	A	H	I	I	J
	Elite	B	C	F	G	H	D	F	G	H	I	A	G	H	I	J	A	B	E	F	G	A	I	I	J	J

Morale Results Key

Result	Infantry and Cavalry	Deployed Artillery	Limbered Artillery	Routers
A	Become disordered, must halt to reform	Become shaken, no movement or firing next turn	Become shaken, no movement or firing next turn	Rally but shaken
B	Become shaken, no movement or firing next turn	Hold orders change to Retire	Hold orders change to Retire	Rally but disordered
C	Assault orders change to Engage	Limber gun and retire 2 inches	Retire 4 inches	Continue Rout
D	Advance orders change to Hold	Limber gun and retire 4 inches	Retire 8 inches	Continue Rout
E	Hold orders change to Retire	Abandon gun and run towards nearest friendly unit	Ride towards nearest friendly unit	Continue Rout
F	Retire 5cm not facing the enemy	Spike gun and retire 6 inches	Retire 4 inches, gun overturned	Continue Rout
G	Retire 10cm not facing the enemy	Crew spike gun and surrender to enemy if within 4 inches	Captured if enemy within 4 inches, rout if not	Continue Rout
H	Retire 6 inches not facing the enemy	Crew surrender if enemy if within 4 inches, gun captured	Rout	Continue Rout
I	Rout as a mob away from the enemy	Crew rout, gun captured	Rout	Break and flee from field
J	Rout scattered and disordered away from the enemy		Rout	Break and flee from field

Definitions

Result	Explanation
Evade	Move away from the enemy at the fastest available rate of movement
Flee	Effectively a rout
Retire	Move away from the enemy facing the threat
Retreat by recoil	Artillery piece is fired and moved back by the force of the recoil and not repositioned in its original position (artillery stand moves back 1 inches)
Scattered	The stands of a unit are dispersed with at least 2 inches between each of the stands
Spike	The crew or the enemy disable an artillery piece
Stalemate	Both sides in a melee have fought one another to a standstill. There is no clear winner or loser in the melee

9. Risk to commanders

Unlike the conflicts of the previous centuries, the Napoleonic Wars did see several commanders becoming casualties. They were often on the battle line leading their troops and sharing the risks. There were also instances of sharpshooters on both sides taking pot shots at these conspicuous figures.

Any commander in a line of battle under fire is therefore at risk. To be in danger, the figures representing the commander should be within 2 inches of a unit under fire, either by musketry or artillery. The firer can nominate one of his firing dice against the commander. Instead of using the standard casualty calculation table, the firer should use the following table:

D10 score	Close Range	Medium Range	Long Range
0-2	Killed outright	Killed outright	Wounded – lose 1 Leadership Point
3-4	Killed outright	Wounded – lose 1 Leadership Point	Stunned – cannot move or issue orders next turn
5-6	Wounded – lose 1 Leadership Point	Stunned – cannot move or issue orders next turn	Horse bolts – move commander 6 inches away from the direct of the fire
7-8	Stunned – cannot move or issue orders next turn	Horse bolts – move commander 6 inches away from the direct of the fire	No effect
9	Horse bolts – move commander 6 inches away from the direct of the fire	No effect	No effect

This table can also be used in cases when a commander comes into contact with an enemy unit. If the commander is contacted in a melee then the attacking player has the option to allocate one of the base casualties he may have inflicted against the commander.

Should the commander be killed, his staff must inform the next commander up the chain of command by courier before a replacement can be allocated. Only a higher commander can make this replacement from his staff.

Only once the courier has reached the superior commander in the command chain can the quality of the replacement be determined; unless named and unattached staff officers have been detailed by the players prior to the battle commencing, this is done by using the Commander Quality Table (see Command and Control, above). Should the unthinkable happen and the army commander is killed or incapacitated and he is above corps level, then use the corps level replacement column in the table.

Replacement cannot actually take place until the figure representing the new commander has

moved to the unit (brigade, division etc) in question. Whilst this is happening, the brigade, division or corps continues to act on existing orders, assuming of course that their morale has withstood the loss of their commander.

10: Objectives and Winning

If you are playing an historical battle, then the objectives should closely follow those of the real battle, although in the majority of cases, beyond either eliminating or routing the enemy army, it is often unclear as to the precise tactical or strategic objectives of the commanders. This is also very much the case on the wargames table. What are the objectives? Is it clear what is needed to win? How long do I have to complete my objectives?

For most wargamers, time (real time) is the deciding factor. These rules are built for speed, so it is unlikely that you will need to decide that the battle will end after X number of turns. A great way of dealing with this problem of deciding who wins and how they win is to assign objectives.

The opposing armies may not necessarily value particular objectives in the same way. If they did, then all units would converge on a particular part of the wargames table and the battle would resemble a rugby scrum. More often than not, beyond the obvious decisive defeat of the enemy (routing all units for example or capturing the grand redoubt and the artillery inside it), real commanders viewed the battlefield in different ways. Certain terrain objectives became an obsession that far outweighed their real worth. To the enemy, the importance of a particular terrain feature eluded them when they were focused on ensuring that a strategic bridge or a road was cleared of the enemy.

Using this premise can produce an exciting and unpredictable game. At times, some terrain features will assume similar levels of importance to both armies, at other times the focus will be entirely different.

We recommend that between 3-5 objectives are agreed before the start of the game. Each objective needs to be given a letter A, B, C, D and E. In secret, each player will roll a D10 and discover the relative importance to them of each of the objectives, write down the value of that objective and only reveal that information when they have achieved their victory point total.

Typically objectives can be roads, bridges, hills or towns and villages. The following tables are designed to help you randomly assign the value of each of the objectives.

3 Objectives priority table

Dice Roll	Priority	Value of Objectives:
0-2	ABC	1st = 10 points
3-4	BCA	2nd = 5 points
5-6	CBA	3rd = 2 points
7-9	ABC	

3 Objectives Victory Table

Objective Points Won	Battle Outcome
17	Crushing Victory
15	Decisive Victory
12	Tactical Victory
10	Marginal Victory

4 Objectives Priority Table

Dice Roll	Priority	Value of Objectives:
0-2	ABCD	1st = 12pts
3-4	BCDA	2nd = 10pts
5-6	CADB	3rd = 5pts
7-9	BCDA	4th = 2pts

4 Objectives Victory Table

Objective Points Won	Battle Outcome
29	Crushing Victory
27	Decisive Victory
22-24	Tactical Victory
15-17	Marginal Victory

5 Objectives Priority Table

Dice Roll	Priority	Value of Objectives:
0-2	ABCDE	1st = 12pts
3-4	DEABC	2nd = 10pts
5-6	EABCD	3rd = 8pts
7-9	DABEC	4th = 5pts
		5th = 2pts

5 Objectives Victory Table

Objective Points Won	Battle Outcome
37	Crushing Victory
35	Decisive Victory
27-34	Tactical Victory
19-26	Marginal Victory

EXAMPLE

Five objectives are chosen by the players. A is a bridge, B is a ford, C a crossroads, D a farm and E a small village. Player A rolls a 6 and the value of his objectives are C(12), D(10), E(8), A(5) and B(2). Player B rolls a 7 and his priorities are D(12), A(10), B(8), E(5) and C(2). The players note down their priorities and the value of the objectives to them and should not reveal this information until the end of the game.

Deciding who has won on this basis is a straightforward matter. Either the possession of sufficient objectives that collectively add up to sufficient points to secure a win, or the possession of objectives after a certain amount of time has elapsed will both work. A reasonable agreement has to be reached with regard to possession of an objective. Simply rushing a group a skirmishers up to the most valuable objective and claiming victory is not sufficient. In reasonable terms the objective has to be secure. It would not be reasonable to have a unit of skirmishers on top of a hill with several enemy regiments bearing down on them and claim that the objective is secure.

If in doubt count up the number of formed units (not counting routers, broken units or skirmishers) within 8 inches of the objective and compare that total with the enemy forces in the same area. If the figure is the same, neither side has secured the objective, if one side has more units than the other, then they have secured the objective.

The proviso with this solution is the question surrounding fortifications and redoubts (and for that matter river crossings and villages). Possession of defensive position is final, but a formed, non-skirmish unit must be holding the objective. It would be unfair to contest the ownership of a redoubt on the other side of a river just because the enemy had managed to manoeuvre more units within the 8 inches range in the last game turn.

Chapter 6

Scenarios

This section comprises of three short scenarios to give you an idea of how to create your own scenarios and how to set up a game. Scenario books accompany this title, the first of which focuses on the Peninsular War and contains a large number of scenarios which you can link together to fight a campaign. We have chosen three scenarios for this section: one from the early period, one from the middle period and one from the Waterloo campaign.

The army lists for these senarios differ from the standard army lists. They are a scaled-down representation of the actual forces involved. There are only two levels of command, so commanders listed as army commanders act as divisional commanders and corps commanders act as brigade commanders in game terms.

Battle of Neerwinden, 18 March, 1793

After the Battle of Valmy the French invaded the Austrian Netherlands. By 6 November 1792 Belgium was under French control and the Austrians retreated to the Rhine. On 18 March 1793 the French attacked the Austrians under Coburg at Neerwinden. The French did not see everything go their own way and the Austrians counterattacked them, driving them back and dispiriting the French army to the extent that a large number of them deserted. The French were forced to retreat and enter into negotiations with the Austrians.

Gaming Neerwinden

The Austrian army begins on high ground overlooking Neerwinden. The key objective is not to gain ground, but to either eliminate or route the enemy forces. The Little Geet River can be forded at half speed if a player wishes.

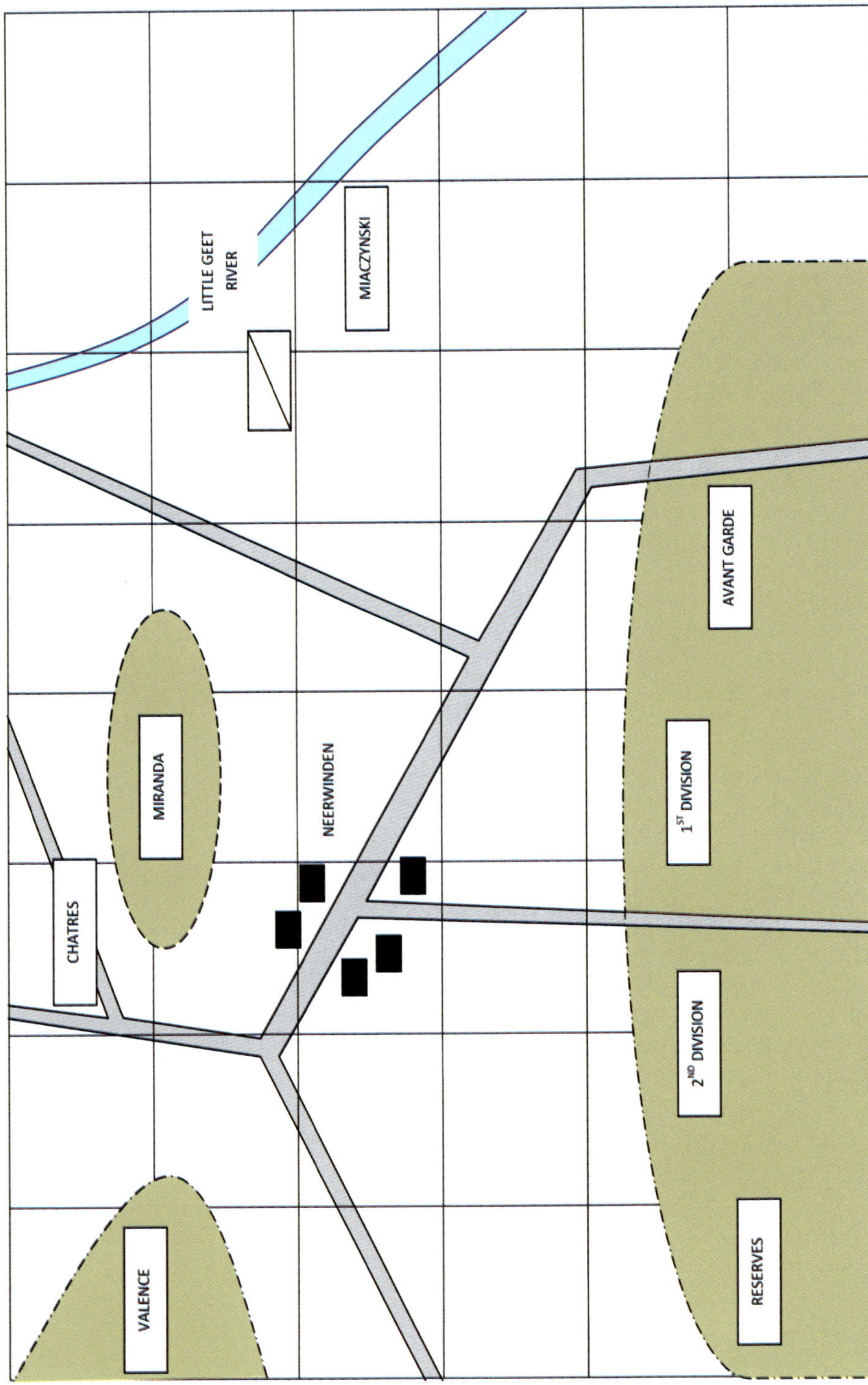

NEERWINDEN March 18 1793, French v Austrians

Austrian order of battle

Army commander: Coburg	-	Able
Avante Garde Commander: Charles	-	Able
1st Line	-	Experienced Steady
2nd Line	-	Experienced Steady
3rd Line	-	Novice Steady
1st Grenadiers	-	Elite Confident
1st Hussars	-	Elite Arrogant
Foot artillery, 2 guns	-	Experienced Steady
1st Division Commander: Ferraris	-	Able
1st Line	-	Novice Steady
2nd Line	-	Novice Steady
3rd Line	-	Novice Steady
1st Cuirassiers	-	Elite Confident
2nd Cuirassiers	-	Elite Confident
Foot artillery 2 guns	-	Experienced Steady
2nd Division Commander: Colloredo	-	Cautious
1st Line	-	Novice Steady
2nd Line	-	Novice Steady
1st Cuirassiers	-	Elite Confident
1st Dragoons	-	Veteran Confident
Foot artillery 2 guns	-	Experienced Steady
Reserve Commander: Clairfayt	-	Able
1st Line	-	Novice Steady
2nd Line	-	Novice Steady
3rd Line	-	Novice Steady
1st Grenadiers	-	Veteran Confident
1st Hussars	-	Veteran Arrogant
1st Chasseurs	-	Experienced Steady
Foot artillery 2 guns	-	Experienced Steady

French order of battle

Army of the North Commander Dumouriez	-	Able
Division Commander: Valence	-	Able
1st Line	-	Novice Steady
2nd Line	-	Novice Steady
1st Grenadiers	-	Experienced Steady
3rd Line	-	Novice Steady
4th Line	-	Novice Steady
5th Line	-	Novice Steady
Chasseurs	-	Novice Steady
Foot artillery 2 medium guns	-	Novice Steady
Horse artillery 2 medium guns	-	Experienced Steady
Division Commander: Chatres	-	Able
1st Line Infantry	-	Novice Steady
2nd Line Infantry	-	Novice Steady
1st Chasseurs	-	Novice Steady
3rd Line Infantry	-	Novice Steady
1st Hussars	-	Novice Steady
Foot artillery 2 medium guns	-	Experienced Steady
Division Commander: Miranda	-	Able
1st Light Regiment	-	Novice Steady
1st Line Regiment	-	Novice Steady
2nd Line Regiment	-	Novice Steady
1st Grenadiers	-	Experienced Arrogant
3rd Line Regiment	-	Novice Steady
1st Hussars	-	Novice Steady
Horse battery, 2 medium guns	-	Experienced Steady
Detachment Commander: Miacynski	-	Able
Hussar Regiment	-	Novice Steady
Chasseur Regiment	-	Novice Steady

Battle of Raab, 14 June 1809

The French commander Eugene was forcing the Austrians out of Northern Italy and into Hungary. However, a setback for the French at Aspern-Essling encouraged the Austrian Archduke John to halt and give battle. The Austrians had around 35,000 men against Eugene's 24,000, supported by MacDonald's 9,000. The French can be considered the aggressor in this battle as the Austrians were far more likely to remain in position and try to hold off the French attacks. In the actual battle the Austrians eventually fled after one of their flanks collapsed. The Austrians retreated towards the fortress at Raab and those that had retreated into the fortifications surrendered a few days later.

Gaming Raab

The river is fordable in front of Mescery's division. If playing on an 8ft table the fordable section should be at least 2ft. If a unit wishes to cross this stretch of the river then throw a D6 and a 5 or 6 discovers a fording point.

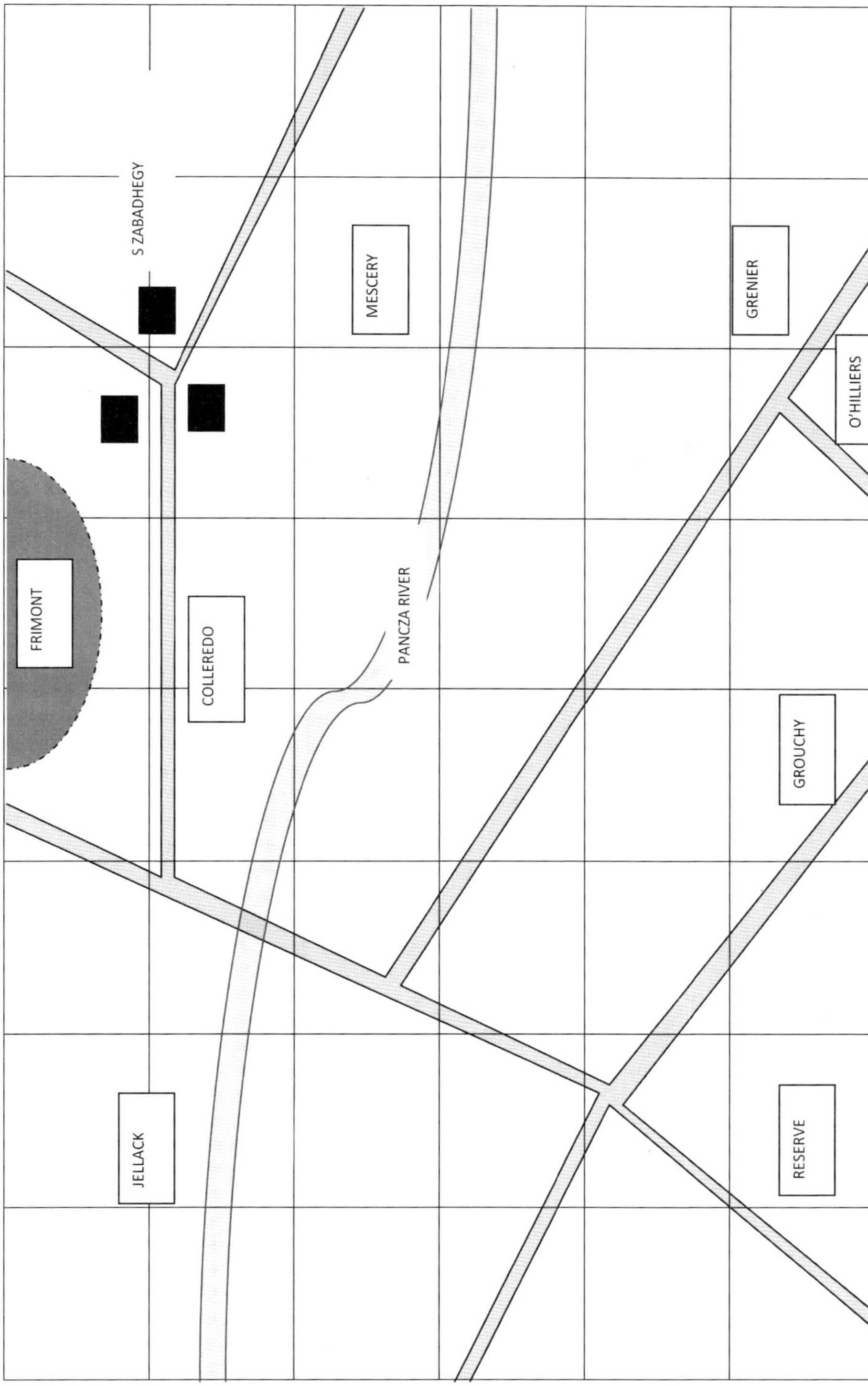
S ZABADHEGY
MESCERY
GRENIER
O'HILLIERS
FRIMONT
COLLEREDO
PANCZA RIVER
GROUCHY
JELLACK
RESERVE

Austrian order of battle

Archduke John	-	Cautious
Left Wing Commander: Mescery	-	Able
1st Hussars	-	Novice Steady
2nd Hussars	-	Novice Steady
Heves and Zemplin Hussars	-	Raw Impetuous
Centre Commander: Colleredo	-	Able
1st Line Infantry	-	Novice Cautious
1st Free Corps	-	Raw Cautious
2nd Free Corps	-	Raw Cautious
Garrison Regiment	-	Novice Cautious
Right Wing Commander: Jellacic	-	Able
1st Line Infantry	-	Novice Cautious
1st Grenzers	-	Experienced Steady
Dragoon Regiment	-	Experienced Steady
2nd Grenzers	-	Novice Steady
1st Hussars	-	Novice Steady
Reserves Commander: Frimont	-	Able
1st Grenadiers	-	Veteran Steady
1st Grenzers	-	Veteran Steady
9th Jagers	-	Experienced Steady
Reserve artillery, 3 medium guns	-	Experienced Steady

French order of battle

Eugene	-	Gifted
Corps Commander: Greniers	-	Able
1st Line Infantry	-	Experienced Steady
2nd Line Infantry	-	Novice Steady
3rd Line Infantry	-	Experienced Confident
Chasseurs	-	Experienced Steady
Corps Commander: De Hillier	-	Able
1st Line Infantry	-	Experienced Steady
2nd Line Infantry	-	Experienced Steady
1st Dragoons	-	Experienced Steady
Reserve Corps Commander: Grouchy	-	Gifted
7th Hussars	-	Elite Arrogant
1st Chasseurs	-	Experienced Cautious
9th Hussars	-	Veteran Confident
1st Dragoons	-	Veteran Confident
Army Reserve Commander: Lecchi	-	Able
Light Infantry Regiment	-	Elite Arrogant
Guard Dragoon Regiment	-	Elite Arrogant
Chasseur Regiment	-	Novice Steady
Baden Jagers	-	Novice Steady
Line Infantry	-	Veteran Confident
Dragoon Regiment	-	Veteran Steady
Artillery battery, 3 medium guns and 1 medium howitzer	-	Experienced Steady

Battle of Ligny, 16 June, 1815

As part of what would become known as the Waterloo Campaign, Napoleon struck at the Prussians at Ligny whilst he detached Ney's smaller force to seize the crossroads at Quatre Bras. Napoleon could not bring his entire army to bear quickly enough at Ligny and by the time he was able to do so night was drawing in and the Prussians were able to slip away. Ligny was Napoleon's last ever victory.

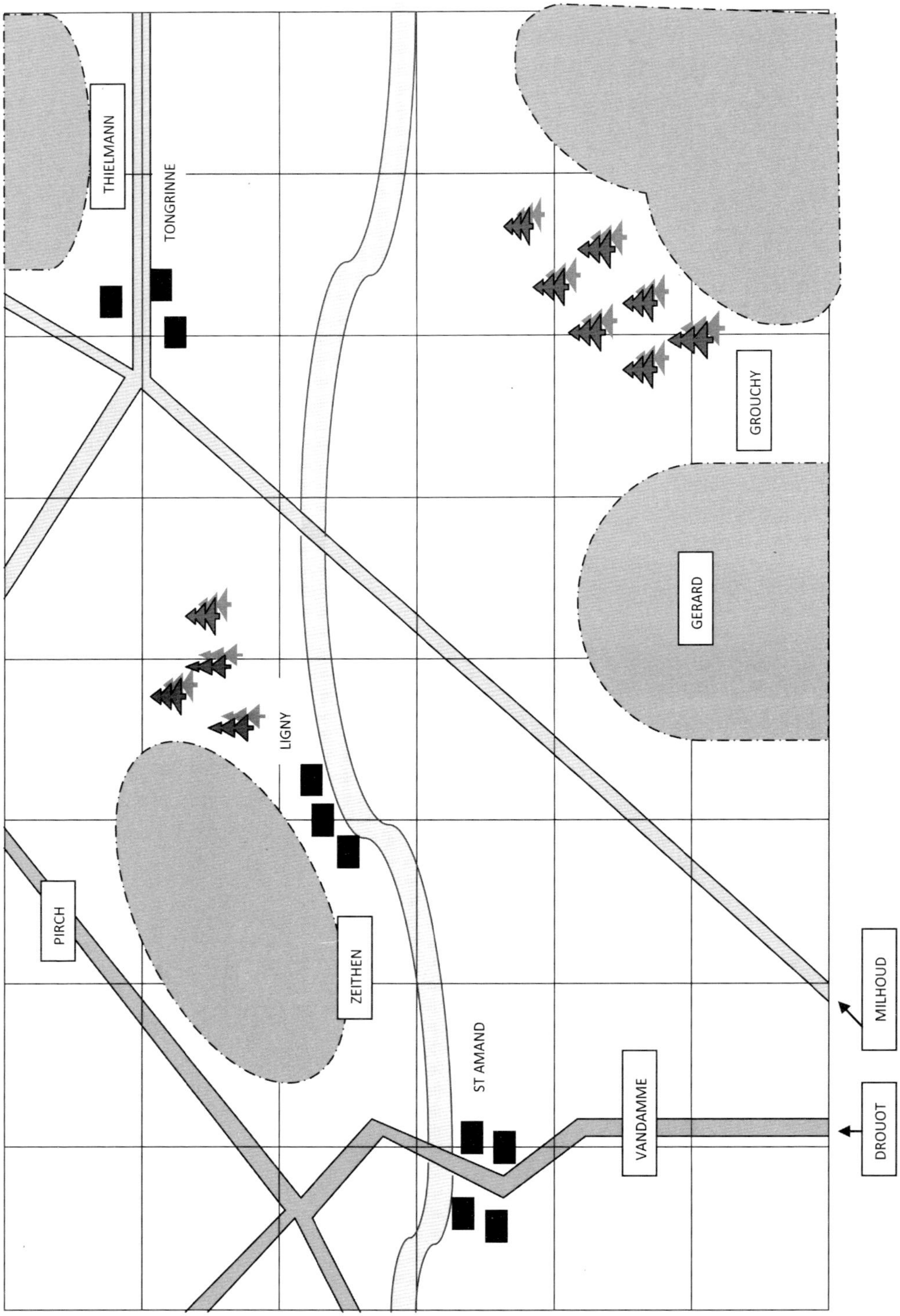

LIGNY June 16 1815, French v Prussians

Gaming Ligny

The river is fordable at almost any point and merely serves to disorganize units crossing it. It therefore only provides the minimum amount of protection from attack.

Prussian order of battle

Blucher	-	Gifted
Corps Commander: Zeithen	-	Gifted
12th Line Infantry	-	Veteran Steady
24th Line Infantry	-	Novice Steady
1st Westphalian Landwehr	-	Raw Cautious
6th Line Infantry	-	Experienced Steady
28th Line Infantry	-	Novice Steady
2nd Westphalian Landwehr	-	Raw Cautious
7th Line Infantry	-	Experienced Steady
29th Line Infantry	-	Novice Steady
3rd Westphalian Landwehr	-	Raw Cautious
Treskow Dragoons	-	Novice Steady
Foot Artillery – 2 medium guns	-	Experienced Steady
Horse Artillery – 2 medium guns	-	Experienced Steady
Corps Commander: Pirch	-	Able
2nd Line Infantry	-	Experienced Steady
25th Line Infantry	-	Novice Steady
4th Westphalian Landwehr	-	Raw Cautious
9th Line Infantry	-	Experienced Steady
26th Line Infantry	-	Novice Steady
1st Elbe Landwehr	-	Raw Cautious
14th Line Infantry	-	Novice Steady
22nd Line Infantry	-	Novice Steady
2nd Elbe Landwehr	-	Raw Cautious
Thumen Dragoons	-	Novice Steady
Hussar Regiment	-	Veteran Impetuous
Foot artillery – 2 medium guns	-	Novice Confident
Horse artillery – 2 light guns	-	Experienced Steady
Corps Commander: Thielmann	-	Able
8th Line Infantry	-	Experienced Steady
30th Line Infantry	-	Novice Steady
1st Kurmark Landwehr	-	Raw Cautious
27th Line Infantry	-	Novice Steady
2nd Kurmark Landwehr	-	Raw Cautious
3rd Kurmark Landwehr	-	Raw Cautious
4th Kurmark Landwehr	-	Raw Cautious
Marwitz Uhlans	-	Veteran Confident
Lottun Dragoons	-	Novice Steady
Foot artillery – 2 medium guns	-	Experienced Steady
Horse artillery – 3 light guns	-	Experienced Confident

French order of battle

Note that Drouot and Milhoud will only arrive on the battlefield on Game Turn 3, at the positions marked on the map.

Napoleon	-	Charismatic
Imperial Guard Commander: Drouot	-	Gifted
Old Guard Grenadiers	-	Elite Arrogant
Middle Guard Grenadiers	-	Elite Arrogant
Old Guard Chasseurs	-	Elite Arrogant
Middle Guard Chasseurs	-	Elite Arrogant
Chartrand Guard	-	Elite Confident
Guard heavy cavalry	-	Elite Arrogant
Foot artillery – 3 medium guns	-	Elite Confident
Horse artillery – 2 medium guns	-	Elite Confident
III Corps, Commander: Vandamme	-	Gifted
8th Infantry	-	Veteran Confident
10th Infantry	-	Experienced Cautious
11th Infantry	-	Experienced Cautious
7th Infantry	-	Veteran Confident
Chasseur Regiment	-	Experienced Steady
Lancer Regiment	-	Veteran Confident
Foot artillery – 2 medium guns	-	Experienced Steady
IV Corps, Commander: Gerard	-	Gifted
12th Infantry	-	Experienced Steady
13th Infantry	-	Experienced Steady
14th Infantry	-	Experienced Steady
Hussar Regiment	-	Veteran Arrogant
Foot artillery – 2 medium guns	-	Experienced Steady
Horse artillery – 2 light guns	-	Experienced Steady
Cavalry Commander: Grouchy	-	Gifted
1st Hussars	-	Elite Impetuous
1st Dragoons	-	Novice Steady
2nd Dragoons	-	Novice Steady
Horse artillery – 3 light guns	-	Experienced Steady
IV Cavalry Corps Commander: Milhoud	-	Gifted
1st Cuirassiers	-	Elite Arrogant
2nd Cuirassiers	-	Elite Arrogant
Horse artillery – 2 light guns	-	Experienced Steady

Chapter 7:

Buying Guide

This buying guide incorporates metal and plastic figures from a wide range of different manufacturers, as well as manufacturers and distributors of buildings and terrain suitable for the Napoleonic period.

One of the first major decisions that has to be made is to the scale of the figures that you will choose to use. Broadly your options are:

- 2mm – a very limited number of manufacturers, but this scale is surprisingly effective and is ideal if space (or budget) is limited.
- 6mm – there is a wide range of sizes that broadly fit into this category, which was formerly known as 1/300th scale. Once you have selected a manufacturer it is unlikely that other ranges will necessarily fit, as the size difference seems more marked in this scale. Again this is an ideal size to go for if your space is limited or you wish to recreate vast battles on a regular size table.
- 10mm – often referred to as the 'grand scale'. This is a relatively new scale for wargaming and is offered by a number of manufacturers. It does combine the ability to have a compact wargames table whilst giving additional scope to paint accurate representations of the units.
- 15-18mm – for many Napoleonic gamers this is the primary scale, as the cost of these figures is around one third that of 25-28mm figures. They allow large units to be built up whilst giving significant scope for painting. The traditional size is 15mm, although many manufacturers now more accurately describe their 15mm ranges as 18mm.
- 20mm – dominated by a large number of plastic figure manufacturers, such as Hat and Italieri. These companies alone provide all the necessary troops you will need to build up an army of almost any nation. There are a number of metal manufacturers that also produce 20mm figures, although due to the differences in casting technique the metal figures are rarely compatible with the thinner and slightly more elegant plastic variety.
- 25-28mm – arguably the broadest possible range of Napoleonic figures offered by a huge number of different manufacturers. The 25mm figures were the original standard size provided by manufacturers such as Garrison and Minifigs. In the 1980s there was a trend towards the larger 28mm figure, such as the Connoisseur and Hinchcliffe ranges. This has now largely become the standard size rather than the more traditional 25mm.
- 40mm – this is a relatively new scale for wargaming in general. However, several manufacturers, including Old Glory and Perry Miniatures, now produce wonderfully detailed figures for this scale. Due to their comparative cost, however, they are probably more suitable for large-scale skirmishes.
- 54mm – there is a mix of plastic, metal and resin figures available in this scale. There are also a number of manufacturers who supply ready-painted figures. Again the general cost for metal and resin figures probably precludes these being used en masse and are probably more suitable for skirmish actions.

Other options include casting your own figures. The Irish-based manufacturer, Prince August, produces a number of suitable moulds, including their extensive Napoleonic range. They produce moulds to cast figures for the 1805 period and for the Battle of Waterloo in 25mm. They also produce 54mm Napoleonic moulds, but this is a more limited range.

This buyers' guide aims to be as comprehensive as possible and cover the majority of manufacturers. Apologies to manufacturers that we may have inadvertently missed out, but that may have a great deal to do with the difficulties in getting in touch with you and finding your website. Many manufacturers assisted in helping us compile this buyers' guide, but others ignored requests for information. It should always be remembered that many manufacturers are part-timers and have other jobs, so do not always expect a swift reply or, in some cases, any reply at all. Always make sure that websites are being updated and maintained and that emails are being answered before placing an order. On the odd occasion, you may happen across an old version of their website or at worst the unscrupulous dealer that has no intention of fulfilling your order.

AB

Eureka Miniatures, 149 Poath Road,
Hughesdale, Victoria 3166, Australia
www.eurekamin.com.au
nicr@eurekamin.com.au
+613956884085

Formerly AB 15mm Napoleonic figure range now available via Eureka Miniatures.

This is a huge range of beautifully-sculpted 15mm Napoleonic figures encompassing Austrians, Bavarians, Brunswickers, Cossacks, French, Grand Duchy of Warsaw, Prussians, Ottomans, Peninsular War British, Portuguese, Russians, Saxons, Spanish and Wurtemburgers

Samples – arguably the finest 15mm figures available on the market. They are beautifully cast, full of detail and character. They are on the smaller, slim side but provide virtually any figure that you could possibly want for Napoleonic gaming.

A Call to Arms

33 Grange Crescent, North Thamesmead,
London, SE28 8EX
www.acalltoarms.co.uk
+44 (0)2082447681

1/72nd (20mm) or 1/32nd (54mm) plastic figures

The 1/72nd scale figures include British infantry and French Cuirassiers, Carabineers and infantry. Infantry are sold in boxes of 32 and cavalry in boxes of 12. The 1/32nd scale figures include French and British infantry, French Cuirassiers, Dragoons and Carabineers, Scots Grayes and British artillery. These are sold as 16 infantry, 8 cavalry or 4 artillery pieces with crew.

Samples – these are polythene figures with good detail. The 1/72nd scale infantry come in 8 poses and are largely flash free.

Accurate (Imex)

PO Box 15390, Brooksville,
Florida 34604, USA
www.imex-model.com
sales@imex-model.com

Formerly a British manufacturer, now owned by Imex. They produce a range of 1/72nd scale plastic figures designed to refight the Battle of Waterloo.

Currently British infantry, cavalry and artillery are available, along with French infantry, cavalry and artillery. They are sold in boxes of approximately 50 infantry and around 20 cavalry. Note that some of their releases are in fact re-boxed Airfix products.

Samples – their originals are clean cast, imaginative poses with most poses being useable.

Adler

129 Bonchurch Road, Brighton,
Sussex BN2 3PJ
www.home.clara.net/adlermin
+44 (0) 1273685279

1/300th scale Napoleonic range. Figures are sold in strips of 4 infantry and 2 cavalry or in battalion packs.

This is a massive range of Napoleonic covering all the major protagonists including the French and the British, as well as less well catered for nationalities, such as the Dutch Belgians.

Samples – Cavalry work out at around 10p per figure and infantry at about 5p per figure. They are cleanly cast with their own definite character.

Assault Group

The Assault Group, Unit K,
Broxtowe Park Business Centre,
Nottingham, NG8 6QP
www.theassaultgroup.co.uk
info@theassaultgroup.com
+44 (0) 1159133591

A small but growing range of 28mm Napoleonic figures depicting the Austro-Hungarian army of 1809.

An interesting new venture for this manufacturer that is more closely associated with WW2 and modern ranges. The figures are full of detail and this promises to be an exciting addition to their current range.

Samples – extremely crisp castings. Infantry sold in packs of 4, currently covering Fusiliers, Grenzerss and Jägers. Unit builder sets are now available containing 36 to 48 figures, which brings the price down to less than £1 per figure.

Baccus

Unit C, Graham House, Bardwell Road,
Sheffield, S3 8AS
www.baccus6mm.com
baccus6@aol.com
+44 (0) 11427224491

A growing range of Napoleonics in 6mm. Infantry are sold in figure packs of 96 or 48 and cavalry in packs of 45. Artillery is sold as 4 guns with crews.

Baccus also sell army packs with sufficient troops to field an entire division. They also sell buildings, bases, flag sheets and other accessories. Currently available are Austrians, Bavarians, British, French, Duchy of Warsaw, Prussians, Russians and Wurtemburgers.

Samples – nicely cast and full of detail, an amazing feat for this scale.

Battle Honours

Units 1 and 2, The Leys, Evesham,
Worcestershire, WR11 2PP
www.battlehonours.co.uk
sales@battlehonours.co.uk
+44 (0) 138645875

Battle Honours offer French revolutionary wars and Napoleonic in 15mm. They are all extensive ranges.

Their French revolutionary wars range includes Austrians, British, French and Russians whilst the Napoleonic range additionally includes Confederation of the Rhine, Polish, Portuguese, Prussian, Saxon and Spanish.

Samples – very nicely cast and delicate 15mm Napoleonic in a very broad range. The only major criticism is the lack of photographs on their website.

Bicorne

The Brown House, Main Street,
Grove, Oxfordshire, OX12 7HX
www.bicorne.net
bicorne@btinternet.com
+44 (0) 1235764018

Bicorne has a number of suitable Napoleonic ranges, including their own original range, the former Connoisseur range, figures that were formerly Firing Line Miniatures and a special campaign series.

The Bicorne Miniatures range covers Austrians, British, French, Swedes and Wurtemburgers. The Connoisseur range is even broader, bringing in additional nations such as the Prussians, Portuguese and Saxons. Added to this are the French and Wurtemburgers available from the former Firing Line Miniatures range. Finally there is the limited but interesting campaign series.

Samples – very competitively priced traditional, larger size 28mm metal figures. Many wargamers are devoted to this scale of figures, which was pioneered by the Connoisseur designer Peter Gilder. The figures hold up very well to modern-day competition.

Dixon Miniatures

Spring Grove Mills, Linthwaite,
Huddersfield, West Yorkshire,
HD7 5QG
www.dixonminiatures.co.uk
dixonminis@aol.com
+44 (0) 1484846162

Range of 25 (28mm) metal figures for the French army in Egypt, French and British in the Peninsular.

On the face of it this is a limited range but the Egyptian campaign figures in particular are virtually unique as they are not covered by many other manufacturers.

Samples – the vast majority of Dixon Miniatures have multiple variations per code, so they have a number of different head variants for each pose, which adds to the character of each unit.

Eagle Figures

46 Bucknall Way, Beckenham,
Kent, BR3 3XN
www.eaglefigures.co.uk
stewart@eaglefigures.co.uk
+44 (0) 208 6588289 (evenings)

25mm British, French, Austro-Hungarian, Russian, Prussian, Spanish and other nations.

In addition to their broad range of Napoleonic for most theatres, the manufacturer also produces Moghul infantry and cavalry for the period, as well as some less well known nations such as the Danes.

Samples – crisply cast, larger size Napoleonic figures full of detail and character and very competitively priced.

Essex	
26 Bowlease Gardens, Bassacarr, Doncaster, South Yorkshire, DN4 6AP www.eliteminiatures.co.uk eliteminiatures@btinternet.com +44 (0) 1302530038	A vast range of 28mm pewter Napoleonic figures encompassing all major protagonists plus Napoleon in Egypt and the excellent collectors' series of French and Austrians.

Elite Miniatures produce one of the best ranges of Napoleonic figures available. The infantry and the cavalry is generally sold in packs of 4, although it is possible to buy whole battalions or regiments as discounted units.

Samples – beautifully detailed, elegant if tall figures. Noticeably flash free and designed to provide a wide variety of different poses and highly animated units.

Elite	
Unit 1, Shannon Centre, Shannon Square, Thames Industrial Estate, Canvey Island, Essex, SS8 0PE www.essexminiatures.co.uk info@essexminiatures.co.uk +44 (0) 1268682309	15mm and 25mm extensive metal Napoleonic ranges

15mm figures are sold in packs of 8 for infantry and 4 for cavalry, although entire armies can be purchased at a discount. The 25mm figures are sold individually or again as ready made armies. The 15mm range covers French, British, Bavarian, Polish, Brunswick, Portuguese, Spanish, Dutch, Austro-Hungarian, Prussian and Russian. The 25mm range covers Russians, French, British and Austro-Hungarian.

Samples – the 15mm range is definitely a question of either love them or loathe them. Some wargamers believe the Essex Miniatures are some of the best Napoleonic available, whilst others describe them as gnomes. Their 25mm figures are well cast and provide the broadest possible range of options.

Eureka	
Eureka Miniatures, 149 Poath Road, Hughesdale, Victoria 3166, Australia www.eurekamin.com.au nicr@eurekamin.com.au +613956884085	Eureka are the manufacturers of 15mm AB range of Napoleonic, but also produce their own range of 18mm figures. They produce British, French and Saxon Napoleonic and are also beginning toproduce a range of early Napoleonic figures.

The castings are exemplary in quality. They are totally flash free and perfect in proportion and detail. There are a number of variants available for each figure. Unique to the wargaming industry, Eureka will also undertake to create any range of figures in any scale, providing a number of wargamers place advance orders to guarantee the initial casting runs and cover sculpting costs. They have just released the first of their 28mm Napoleonic.

Samples – excellent detail, beautifully cast and very affordable. Available through agents, as well as direct.

Fantassin (Warmodelling)	
Avenida de Astorga, KM1, 24400 Ponferrada, Leon, Spain www.warmodelling.com info@warmodelling.com +34 987428602	An absolutely enormous range of 15mm Napoleonic figures covering Austrians, French, Prussians, Swedes, Bavarians, Peninsular War British, Russians, Grand Duchy of Warsaw, Portugal and Spain

This is a truly vast range and as comprehensive as anything available on the market. Worldwide they are relative newcomers and the figures are extremely competitively priced. However they are priced in Euros. Infantry figures are generally sold in packs of 8 and cavalry in packs of 4. There is a massive range of equipment and accessories.

Samples – the figures are also available in battalion bags, which make the figures even more affordable. More recently it is now possible to buy division bags of infantry and cavalry. The figures are beautifully detailed, very easy to paint and very much designed with the wargamer in mind, with a number of different poses within each pack code.

Flashing Blade

37 Oldbarn Close, Southampton,
SO40 2SY
www.flashing-blade.com
sales@flashing-blade.com
+44 (0) 2380861397

A growing range of 10mm British, French, Brunswick and Prussian figures.

The figures are sold in relatively large packs of 60-100 infantry, 9 guns with 36 crew or 24 cavalry. Currently there are enough different figure variants to be able to produce credible British and French units. The other nationalities have limited ranges at the moment.

Samples – the figures are nicely cast, delicate and look fantastic en masse. They have simple-to-paint detail.

Foundry

23-34 St Mark's Street,
Nottingham, NG3 1DE
www.wargamesfoundry.com
sales@wargamesfoundry.com
+44 (0) 1158413000

Arguably one of the largest ranges of Napoleonic available in 28mm metal.

Rightly or wrongly, the Foundry has come under criticism for being one of the first manufacturers to adopt the pack as the primary way of being able to purchase figures. Their packaging policy at the moment is somewhat confusing, with some packs containing 8 figures whilst others only have 6. All, however, are the same price. Currently, Foundry offers a broad range of the main nationalities, although some of their ranges go in and out of production with no warning.

Samples – Foundry figures are wonderful to paint. They are designed by wargamers for wargamers. Many of the figures are produced by some of the foremost sculptors in the world. You can get relative bargains if you take advantage of their periodic sales and purchase their collections, which are post free. Otherwise postage is comparatively high if you just purchase a handful of packs.

Front Rank

The Granary, Banbury Road,
Lower Boddington, Daventry,
Northamptonshire, NN11 6XY
www.frontrank.com
info@frontrank.com
+44 (0) 1327262720

An extensive range of 28mm Napoleonic, including Austrians, Bavarians, British, French, Nassau, Dutch, Polish, Portuguese, Russian and Spanish.

The figure ranges for each of the nationalities is comprehensive and complete, including infantry, cavalry, generals, artillery crews and good value battalion packs, which would be ideal to represent a regiment. Their figures are full of characters, are of average price and very easy to paint.

Samples – their figures are very crisp castings, with no flash or miscasts. The ranges are continually under review and improvement. The battalion packs are particularly good value as each infantry figure works out at slightly less than £1.

Garrison

16 Market Lane, Walpole St Andrew,
Wisbech, Cambridgeshire, PE14 7LT
www.garrisonminiatures.com
info@garrisonminiatures.com

Garrison have been producing 25mm figures since 1966. Their Napoleonic range covers Austrians, British and French.

In some people's eyes these figures may be considered somewhat rough and ready by modern day standards. Comparatively speaking they are, but they actually paint up very well and they are less than half the price of most contemporary competitors. In fact buying in packs of 10 means that the figures are just 50p each.

Samples – when samples arrived we were surprised to see just how clean and well detailed the figures still were after having purchased some of them nearly 30 years ago. They are definitely 25mm figures and are therefore compatible with more traditional ranges, such as Minifigs and would look out of place with modern day 28mm figures.

Grand Manner

Unit B, Smith's Way, Saxon Business
Park, Stoke Prior, Bromsgrove,
Worcestershire, B60 4AD
www.grandmanor.co.uk
info@grandmanor.co.uk
+44 (0) 1527578766

28mm Latin and Hispanic resin buildings

Although Grand Manner does not manufacture figures, they do produce a fantastic range of 28mm resin buildings. They are not cheap, but they would provide an incredible centrepiece to any Peninsular War battle.

Samples – faultless mouldings with unbelievable detail. The only problem you may have with them is being able to do justice to the detail with your painting. All pieces are available painted, for example at the time of writing, a plain Hispanic chapel with bell tower was £54 unpainted and £162 painted to a collector's standard.

Hat

PO Box 50284, Bellevue
WA 98015-0284, USA
www.hat.com

An enormous range of 1/72nd scale plastic figures covering French, Bavarian, Swedish, Prussian, Russian, British and allied, and Austrian, Wurtemburg, Saxon and Spanish.

If you choose to take the route of 20mm plastic figures then the Hat range probably will give you everything that you will need. The range is constantly growing with, on average, 5 or 6 new sets being added to it each year. The infantry are generally sold in boxes of 48 or 100 whilst the cavalry are sold in boxes of 12. Also available are a growing number of different wagons for the period. Note that some of the figures available from Hat are Airfix reissues, such as the classic Waterloo British and French infantry and the Prussian Landwehr. These tend to go in and out of stock but are generally available from other sources.

Samples – it would be impossible to try to do justice to the entire range by giving a blanket review of their figures. However, increasingly the figures are being cast in a slightly more pliable but resilient plastic. This takes paint far more easily. Arguably, their Napoleonics are the best of the ranges that they offer. As these new figures are designed for wargamers there are virtually no useless and unusable poses in the boxes any more.

Heroics & Ros

Heroics & Ros,

3 Castle Way,
Feltham, Middlesex, TW13 7NW
www.heroicsandros.co.uk
sales@heroicsandros.co.uk
Under the new ownership of Andy Kirk

Wide 1/300th scale range

including all the major powers, plus Spanish, Portuguese, Sweedish, Bavarian and Duchy of Warsaw.

The 1/300th scale figures are sold in packs of around 48 infantry, 20 cavalry, 3 horse-drawn guns or 6 foot guns. They represent exrtremely good value for money and were the very first 6mm figures to be produced.

Hinchcliffe

Hinds Figures Ltd, 99 Birchover Way,
Allestree, Derby, DE22 2QH
www.hinchcliffe.co.uk
enquiries@hinchcliffe.co.uk
+44 (0) 1332559025

Hinchcliffe has been in production since the 1970s and Hinds Figures are the latest owners of the range. Along with Minifigs this was the mainstay of 25mm Napoleonic wargaming. The range appears to be gradually coming back into production, incorporating the still wonderful Foremost range, which produced stunning Highlanders and French cavalry.

Although it will take some time for Hinds to bring the whole range back into production, they are making enormous steps and the bulk of the Prussians and French and selected British are already available at the time of writing.

Samples – you may find the standard Hinchcliffe figures to be rather crude or rough and ready compared to modern day equivalents. Hinds have kept the price under £1 per figure, but this may be a problem and may dissuade new gamers to the period, but it will over-joy existing Napoleonic wargamers that want to finish off regiments and divisions.

Irregular

41 Lesley Avenue, York,
YO10 4JR
www.irregularminiatures.co.uk
email@irregularmin.fsnet.co.uk
+44 (0) 1904671101

2mm, 6mm, 10mm, 15mm, 20mm, 25mm, 42mm, 54mm Napoleonic ranges.

Irregular Miniatures offer an absolutely enormous number of options for the Napoleonic gamer, from the diminutive 2mm to the skirmish size 54mm figures. Irregular Miniatures' figures are somewhat variable in quality and some ranges have been around for many years. The 2mm figures are surprisingly good and figures are sold in blocks of ranks of up to around 48 figures. The more common scales such as 15mm offer most of the main protagonists, including French, Austrian and British. Their 25mm range is on the smaller side of 25mm and at present only includes British and French.

Samples – their figures are well cast and lifelike. The figures are competitively priced, very easy to paint and if bought in bargain packs the 25mm figures are only 60p each.

Italieri

Via Pradazzo, 61-40012,
Calderara di Reno, Bologna, Italy
www.italieri.com
+39 (051) 3175211

1/72nd scale plastic figure manufacturer offering French, Prussian, British and Russian infantry, cavalry and artillery

There is a broad range of Italieri figures available of variable quality and usefulness of pose. Broadly, infantry are sold in boxes of 50 and cavalry in boxes of 18. You may find some rather more fanciful poses in these boxes compared to Hat.

Samples – soft plastic, which after undercoating takes paint extremely well. They are well-detailed, crisp castings with very much a style of their own. Many of the figures match well with their Russian sometime collaborators, Zvezda.

Lancashire

28 Helmshore Avenue, Moorside,
Oldham, OL4 2QB
www.test.gtns.co.uk
lancgames@lineone.net

A large range of French, Russian, Spanish, Portuguese, Austrian, Grand Duchy of Warsaw, Bavarian, Prussian and British and their allies in 15mm

The figures are sold in packs of 60 foot or 24 mounted. Currently there are around 100 codes available, although the manufacturer only produces limited artillery.

Samples – these are on the smaller scale of 15mm but actually represent very good value for money, as the infantry figures are just 18p each.

Magister Militum

Unit 4, The Business Centre,
Morgan's Vale Road, Redlynch,
Salisbury, SP5 2HA
www.magistermilitum.com
info@magistermilitum.com
+44 (0) 1725510110

10mm Napoleonic figures, currently offering French and Russian ranges.

Although this is a relatively limited range at present, the most French and Russian units are offered. Figures are sold in either packs of 30 infantry or 12 cavalry.

Samples – nice crisp little castings, full of detail and character; some of the best available in this scale.

Minifigs

Caliver book

100 Baker Street,
Newthorpe, NG16 2DP
www.miniaturefigurines.co.uk
minifigs@hotmail.co.uk
+44 (0) 01159382111

15mm and 25mm Napoleonic ranges, including 15mm French Revolutionary period figures

Minifigs are one of the older businesses still in operation and offer a vast range of 15mm Napoleonic figures, incorporating almost every nationality. Each of the ranges is extremely comprehensive, covering not only each troop type but also each sub-period within the Napoleonic Wars. 15mm infantry tend to be sold in bags of 8 and 25mm infantry are sold singly at well under £1 each. Acquired by Caliver Books in 2009.

Samples – some of these figures have been in production for 30 years and to be honest some of them do show signs of that. But Minifigs have ensured that the quality of the casting remains high and they are not using ancient production moulds. They are on the small side of 25mm and of 15mm but provide enough variation to only collect from this manufacturer if desired.

Museum

28 Long Lane, Driffield,
Yorkshire, YO25 5HF
www.museumminiatures.co.uk
dave@museumminiatures.co.uk
+44 (0) 1377241010

15mm British, French and Prussian Napoleonic figures plus Napoleonic army packs

Infantry sold in packs of 8, cavalry in packs of 4 and command packs with 6 foot or 3 cavalry. A relatively limited range, but very competitively priced.

Samples – nice crisp castings. The British focus primarily on 1812, so are ideal for the Peninsular War. The same can be said for the French. Unlike many manufacturers' websites, all packs have clickable codes, which show a photograph of a typical figure.

Navwar

11 Electric Parade, Seven King's Road,
Seven Kings, Ilford, Essex, IG3 8BY
www.navwar.co.uk
+44 (0) 2085906731

A rare beast indeed as this is a wargames shop just outside of London. They are stockists of 15mm Napoleonics.

The Naismith Napoleonics still hold up extremely well even though they were designed some years ago. They include Austro-Hungarians, Bavarians, Dutch Belgians, French, British, Russian, and Spanish and Swedish infantry. There are also Turkish troops available.

Samples – the 1/300th scale figures are extremely tiny and delicate compared to other manufacturers such as Baccus. Nonetheless they are extremely usable and paintable figures. Being able to buy an entire regiment for £2.50 has to be a bonus.

Newline

32 Royal Crescent, South Ruislip,
Middlesex, HA4 0PW
www.newlinedesigns.co.uk
newlinedesigns@aol.com
+44 (0) 2088456643

Large Napoleonic range in 20mm metal and a smaller 25mm range.

A very comprehensive 20mm range of figures, some of which are broadly compatible with 1/72nd scale plastics. Most nationalities are available, including personalities and naval figures. In 25mm the range is smaller but covers British, Brunswick, French, Nassau, Prussians, Russians, Portuguese, Dutch, Belgians, Wurtemburgers, Bavarians and Spanish. Starter and bargain packs, as well as unit packs, are also available.

Samples – chunky 20mm and short 25mm figures that look great when they are painted up, although in bare metal look rather flat in detail.

Odemars (Ykreol)

www.ykreol.com	Interesting if crude 1/72nd scale Spanish, Portuguese and French Napoleonic

An odd company that produces just a handful of sets with probably the worst box art you are likely to see. But they do however produce a wonderful plastic Hougoumont for the battle of Waterloo. All of the buildings are available including the gardener's house and cow shed.

Samples – as already mentioned the figures are rather crude yet they do Wellington's staff consisting of 42 figures, although you might wonder why you would want 4 Wellingtons in one box. Up and coming are French and British infantry to be used with their Hougoumont buildings.

Old Glory

Old Glory Corporation, Institute House, New Kyo, Stanley, County Durham, DH9 7TJ www.oldgloryuk.com info@oldgloryuk.com+ 44 (0) 1207283332 Note for US customers www.oldglorymininiatures.com	A vast range of grand scale 10mm Napoleonic, 25mm Napoleonic, Sash and Sabre 25mm and a growing range of Sash and Sabre 40mm Napoleonic

The grand scale 10mm Napoleonic range incorporates a bewildering range of French, Austrian, Russian, Prussian, British and French allies. Infantry figures are sold in packs of 50 and cavalry in packs of 30. Virtually nothing has been left out of this range. The 25mm Napoleonic are equally as impressive with French, Russian, Prussian, Austrian, British, Polish and French in Egypt. Infantry are sold in bags of 30 and cavalry in bags of 10. The Sash and Sabre range is primarily French, Austrian and Russian. The growing Sash and Sabre 40mm range offers the most affordable way of getting into 40mm Napoleonic, with 5 infantry figures for £9.

Samples – excellent detail, enormous range and highly competitive prices make Old Glory one of the major Napoleonic manufacturers. Added to this is the variety of scales available to suit any pocket or available space. Within each pack there are a number of variant figures. This is certainly one of the authors' preferred suppliers.

Pendraken

1 Easby Grove, Eston, Middlesbrough, North Yorkshire, TS6 9DL www.pendraken.co.uk d.pengille@btinternet.com +44 (0) 1642460638	10mm Napoleonic for British, Austrian, French and for the 1812 campaign.

Infantry are sold in packs of 30 foot, cavalry in 15s and 2 guns with limbers, horses and crews. Army packs are also available.

Samples – nicely detailed figures with a nice variety of poses and troop types. At present the Moscow campaign Napoleonic range offer nice multi-pose figures in campaign uniforms.

Perry Miniatures

PO Box 6512, Nottingham, NG7 1UJ www.perry-miniatures.comperry miniatures@aol.com +44 (0) 1159503242	A range of 25mm Brunswick and Dutch Belgians, 40mm Peninsular War and forthcoming 28mm plastic French and British for the 1812-1815 period.

For many Perry 28mm and 40mm figures are the pinnacle of modern design. Each pack contains a variety of different poses. They are comparatively well priced for their quality and attention to detail. The 28mm range also includes Nassau figures. The growing Peninsular War 40mm range is impressive, although the figures are comparatively expensive.

Samples – Perry Miniatures are second to none in terms of attention to detail, crisp castings and determination to complete a range, which is often missed by many other manufacturers. The arrival of their 28mm plastic figures should revolutionise 28mm Napoleonic wargaming.

Prince August

Prince August Ltd, Kilnamartyar Village,
Macroom, County Cork,
Republic of Ireland.
www.princeaugust.ie
info@princeaugust.ie
+353 (0) 2640222

A manufacturer of 25mm and 54mm moulds for home casting.

At present Prince August offers a range of moulds for 1805 and a second set suitable for troops to fight Waterloo. Each of the infantry moulds allows you to cast 3 figures at a time. The 54mm moulds are triple figure castings and could be described as borderline toy soldier style.

Samples – n/a

Revell

Henschelstrasse 20-30, D-32257
Bunde, Germany
www.revell.de
contact_d@revell.de
+49 (0) 52239650

A range of 1/72nd plastic Napoleonic

One of the problems with Revell is that a number of the figure boxes periodically go in and out of production without any apparent warning or reason. Infantry boxes contain around 40 figures, cavalry (bearing in mind some of these are Italieri figures) with around 18 figures and artillery (also Italieri) with 2 guns, limber and crews.

Samples – delicate, bendy plastic figures. Their own figures are superior to the Italieri figures they produce under license. If you see a range you want, such as Napoleonic British infantry then buy as many boxes as you need, otherwise you may find you won't be able to get them for 3 or 4 years.

Sash and Sabre

See Old Glory

Spencer Smith

Peter Johnstone, The Old Rectory,
Wrotham, Diss, Norfolk, IP22 1SL
www.spencersmithminiatures.co.uk
peter@spencersmithminiatures.co.uk
+44 (0) 1379650021

Manufacturers of Willie 30mm early Napoleonic and later Napoleonic, and Tradition 30mm Napoleonic

Spencer Smith Miniatures has become the home of several venerable manufacturers dating back to the 1960s. The Willie figures have their devotees, as do the beautifully detailed and delicate Tradition Napoleonic figures (the authors own several hundred Tradition ACW figures).

Samples – whilst the old-fashioned style may not necessarily appeal to everyone, many wargamers from the 60s and 70s had no other option but to use these figures or Airfix in order to game the Napoleonic period. They are comparatively expensive at just over £1 per figure.

Strelets

www.strelets-r.com	A range of 1/72nd scale plastic figures with a definite bent towards the Russian campaign

An interesting variety of Napoleonic French, Russian and Prussian figure boxes. Some other useful packs including allied chiefs of staff, which include all generals and corps commanders. They also do Russian and Prussian chiefs of staff.

Samples – generally available through many hobby stores and distributors, the figures are cruder than Hat or Italieri figures, but have their own distinct charm. The campaign series is particularly good with the same types of figures done in both summer and winter dress.

Time Cast

GDS Building, Lancaster Road, Shrewsbury, SY1 3NF www.timecastmodels.co.uk sales@timecastmodels.co.uk +44 (0) 1743465676	6mm, 10mm and 15mm resin cast buildings and scenic accessories

Very nicely cast range of buildings to suit both southern and western European theatres of war. The 15mm range is rather more orientated towards the 20th century but there are specific ranges for Leipzig and Borodino, as well as Waterloo in both 6mm and 10mm.

Samples – n/a

Victrix

Victrix Ltd, 60 Elsenham Street, London, SW18 5NT www.victrixlimited.com sales@victrixlimited.com +44 (0) 2088741838	28mm plastic Napoleonic British and French infantry

At the time of writing there are eight boxes of multi-part 28mm figures available. All are infantry boxes with upto sixty figures each. Cavalry are promised in the near future. The range is supported by metal specialists and commanders.

Samples – on the large side, but very affordable, adaptable and cleanly cast.

Warrior

14 Tiverton Avenue, Glasgow, G32 9NX www.warrioronline.demon.co.uk info@warriorminiatures.com +44 (0) 1417783426	15mm and 25mm Napoleonic figures

Some of these figures have been around for many years and they are definitely on the smaller side in each case. They are, however, one of the cheapest metal manufacturers in the world and if full armies are purchased then the price includes free shipping to anywhere in the world. Currently there are British, French, Bavarians and Austrians in 15mm and British, French, Austrian, Russian, Spanish and Prussian available in 25mm.

Samples – many people will not like Warrior Miniatures figures, particularly the 25mm ranges which are very spindly and crude. The 15mm figures are nicer and better proportioned.

Waterloo 1815	
www.hatitalywaterloo.it +390586864596	Italian distributors of most plastic manufacturers but they also produce their own 1/72nd scale figures

At present Waterloo 1815 has a limited range of appropriate figures of their own, but these include Polish infantry, Italian carabineers, Belgian artillery and Austrian infantry and artillery.

Samples – to look at the figures are very similar to Esci (now available via Italieri) and presumably they use the same sculptor. Their Austrians are actually designed for a later period. They are fairly standard 1/72nd scale plastic figures.

Zvezda	
141730 Lobnya, Moscow Reg, Promyshlenaya Street 2, Russia www.zvezda.org.ru zvezda@df.ru +7 (495) 5771163	1/72nd scale plastic Napoleonic; primarily French and Russian

As far as plastic figures are concerned these are probably the pinnacle in terms of design. Some of the sets are in fact re-boxed Italieri products. Infantry are generally sold in boxes of around 40 and cavalry in boxes of 16 to 18. The artillery is especially good, as they incorporate limbers and wagons with full complements of horses..

Samples – exquisite figures, particularly the manufacturer's own creations. Their Russian cuirassiers are probably the finest 1/72nd scale plastic figures available.

Space does not allow us to look in any great detail at the wider range of generic wargames manufacturers that offer terrain items or accessories, although many of these manufacturers offer links to (or offer themselves) buildings, bases, paints, brushes, terrain boards and a host of other vital equipment for the wargamer.

Bibliography and Further Reading

Cassidy, Martin, *Marching with Wellington* (Pen & Sword Books, 2003)

Chandler, David, *Napoleon* (Pen & Sword Books, 2001)

Dempsey, Guy, *Albuera 1811* (Frontline Books, 2008)

Dodge, Theodore, *Napoleon's Invasion of Russia* (Frontline Books, 2008)

Field, Andrew, *Talavera* (Pen & Sword Books, 2005)

Gill, John H, *Thunder on the Danube* (Frontline Books, 2008)

Gillespie-Payne, Jonathan, *Waterloo*(Pen & Sword Books, 2004)

Glover, Michael, *Warfare in the Age of Bonaparte* (Pen & Sword Books, 2003)

Haythornthwaite, Philip, *The Waterloo Armies* (Pen & Sword, 2007)

Hofschroer, Peter, *Waterloo 1815, Quatre Bras and Ligny* (Pen & Sword Books, 2005)

Hopton, Richard, *The Battle of Maida 1806* (Pen & Sword Books, 2002)

McGuigan, Ron, Howie Muir and Roy Muir, *Inside Wellington's Peninsular Army* (Pen & Sword Books, 2006)

Mikaberidze, Alexander, *The Battle of Borodino* (Pen & Sword Books, 2007)

Paget, Julian, *Wellington's Peninsular* (Pen & Sword Books, 2005)

Richards, Donald S, *The Peninsular Years* (Pen & Sword Books, 2002)

Robertson, Ian, *Wellington At War In The Peninsula* (Pen & Sword Books, 2000)

Rudorff, Raymond, *War to the Death* (Pen & Sword Books, 2005)

Smith, Digby, *Napoleon Against Russia* (Pen & Sword Books, 2005)

de Tascher, Maurice, *Campaigning for Napoleon* (Pen & Sword Books, 2006)

Quick Reference Sheets and Order Markers

Upon purchase of this book, permission is granted to photocopy or scan the sheets in the following section for personal use. Alternatively they can be downloaded from the Pen & Sword website: www.pen-and-sword.co.uk

Quick Reference Sheet - Game Turn and Command

Game Turn Sequence (p.134)

1. Brigade commanders issue orders or change orders to regiments
2. Brigade commander sends messenger to Divisional Commander if change of orders required
3. The C-in-C issues new orders or makes a 'lucky guess'
4. Insert Random Event card if scenario requires
5. Draw first Brigade Card (see Card Deck)
6. Challenge Initiative (see Challenge Initiative)
7. Player with initiative takes his turn
8. Next card drawn, player takes turn; continue till all Brigades have taken action
9. Morale Checks
10. Compulsory moves
11. Shuffle Brigade card deck, ready for next turn

Orders to Corps or Divisions (p.130)

Seize: At least 1 subunit must be given seize orders, all others either support or manoeuvre
Support: Specify formation to support. No subunit may be given seize, hold or fight. with.
Hold: At least 1 subunit must be given hold orders, none may have seize or fight. with.
Manouevre: Specify direction. At least half of subunits must have manoeuvre orders
Fighting Withdrawal: At least one subunit must be given fighting withdrawal orders

Orders to Brigades (p.130)

Seize: At least 1 regiment must be given assault orders, others engage or advance
Support: No regiments may be given assault, hold or retire orders
Hold: At least 1 regt must be given hold orders, none may be given advance or assault
Manouevre: At least half regiments must have advance orders
Fighting Withdrawal: At least one regiment must be given retire order.

Orders to Regiments (p.132)

Order	Purpose and limitations of the order
Assault	Advance aggressively towards the enemy, charge or attack, fight or shoot
Advance	Move forward towards the enemy, fight or shoot
Engage	Move as required, fight or shoot
Hold	Defend a position or objective, fight or shoot, no movement
Retire	Move away from the enemy, fight or shoot if attacked

Command Radius (inches): Poltroon 8, Cautious 8, Able 12, Gifted 16, Charismatic 16
Units under command: Poltroon 2, Cautious 3, Able 4, Gifted 6, Charismatic 8
For commander casualties see pp157 and for commander replacement see pp128

Out of Command Table

	D10 score				
Unit's training	**0-1**	**2-3**	**4-5**	**6-7**	**8-9**
Green	A	C	D	D	E
Novice	A	B	C	D	E
Experienced	A	C	E	F	F
Veteran	A	B	C	E	F
Elite	A	A	B	C	E

A - Obey orders as given. **B** - Carry out previous orders or result. **C** - Copy neighbouring unit within 4", otherwise hold.
D - Advance 4" if possible, otherwise hold. **E** - Retire 4". **F** - Hold in place)

Quick Reference Sheet – Movement

Movement in inches:		Road/ Track	Open	Rough/ Ford	Dense Wood	Wood	Swamp /Marsh	Steep hill/ slope
Infantry	Column	10)	7	5	4	5	3	4
	Line	4	4	3	2	3	2	2
	Extended	5	5	4	3	4	3	3
Cavalry	Column	20	12	8	▲	8	•	6
	Line	8	8	6	▲	6	•	4
	Dismounted	4	4	3	2	3	2	2
Artillery	Limbered	12	10	6	■	6	■	2
	Manhandled	3	2	2	■	1	■	1
	Skirmishers	8	8	6	4	6	4	4
Couriers		20	20	20	20	20	20	20

▲ - Cavalry must dismount to pass through and move at dismounted cavalry rates. • - Impassable to cavalry, must go round the obstacle. ■ - Impassable to artillery, must go around the terrain feature

Double Time Movement

Troop type	Road	Open	Combination of Road and Open
Infantry	14	10	12
Cavalry	28	20	24
Artillery	18	14	16

Only available to limbered artillery or units in column and on road or in open.

Double time must not come within 10 in of enemy cav (unless friends intervening) or 4 inches of enemy in any circumstances. Must have 1 unit of normal movement between double time moves.

Changes of direction:

Oblique upto 450 of straight ahead, keep original facing. Change facing by wheeling, measure from outside corner: Exp, Vet and Elite units wheel at full speed, Green and Trained units lose 1/4 move unless in Extended line. Skirmishers move and face any direction so long as no base exceeds move allowance.

Formation changes:

Infantry, column > line > extended line > skirmish order (or reverse), 1/2 move each step.
Cavalry, column > line (or reverse), mount or dismount, 1/2 move each
Infantry may form square from line or column (or reverse): Green 2 turns, Trained 11/2 turns, Experienced 1 turn, Veteran or Elite 1/2 turn

Charge Movement Table

Formation	Road/Track/Open	Rough	Woods	Downhill	Steep Slope
Inf Line	6	5	5	6	3
Inf Extended	7	6	6	7	4
Inf Col *	7	5	3	6	4
Cav Line	12	8	8	10	6
Cav Column	14	7	7	14	7

*Only French infantry may charge in column.

Quick Reference Sheet - Firing (1)

1. Check range: short (s), medium (m) or long (l)
2. Determine number of firing dice (min. 1)
3. Roll dice, select best score(s), refer to Musketry or Artillery Fire Table as approriate
4. Check meaning of result against Firing Results Key
5. Conduct any morale checks specified on Firing Results Key

Weapon Ranges (in inches)

Small Arms	**Short**	**Medium**	**Long**
Pistol	0 - 1	over 1 - 2	over 2 - 3
Cavalry Carbine	0 - 3	over 3- 5	over 5 - 7
Musket	0 - 4	over 4 - 6	over 6 - 8
Rifle	0 - 5	over 5 - 8	over 8 - 12

MUSKETRY FIRING TABLE

range Firing Unit	Target & (s,m, l) best roll	Inf. line S M L	Inf. Col square S M L	Inf ext line S M L	Skirmishers* S M L	Mtd cavalry S M L	Unlim artill'y S M L	Limb'd artill'y S M L
Green	0-2	A B B	A A B	B B C	B B D	A A B	A B C	A C D
(1D10	3-4	B B C	A B C	C D E	B C F	A B C	B C D	B D E
before	5-6	C D D	C C E	D E F	C D F	B C D	C D E	C E F
modifiers)	7-8	D E F	D E F	E E F	D E F	C D E	E F F	D F F
	9	E F F	E F F	F F F	E F F	D E F	F F F	E F F
Trained	0-2	A A B	A A B	A B C	A B C	A A B	A A B	A B C
(2D10	3-4	B B C	A B C	B C D	B C D	A B C	A B C	B C D
before	5-6	B C D	B C D	C D E	C D E	B C E	B C D	C D E
modifiers)	7-8	C D E	C D E	D E F	D E F	C D E	C D E	D E F
	9	D E F	D E F	E F F	E F F	D E F	D E F	E F F
Experienced	0-2	A A B	A A B	A A B	A B C	A A A	A A A	A B B
(3D10	3-4	A B C	A B B	B B C	A C D	A B B	A B B	A B C
before	5-6	B B D	A B C	C D E	B D E	B C C	B C C	B C D
modifiers)	7-8	C C E	B C E	D E F	C E F	C D E	C D E	C D E
	9	D E F	C D F	E F F	D F F	D E F	D E F	D E F
Veteran	0-2	A A B	A A A	A A B	A A B	A A A	A A A	A A B
(4D10 before	3-4	A A B	A A B	A B C	A B C	A A B	A A B	A B C
modifiers)	5-6	B B C	A B C	B C D	B C D	B B C	A B C	B C D
	7-8	B C E	B B E	C D E	C E F	B C D	B C D	C D E
	9	C D F	C C F	D E F	D F F	C D E	C D E	D E F
Elite	0-2	A A A	A A A	A A A	A A A	A A A	A A A	A A A
(5D10 before	3-4	A A B	A A B	A B B	A B B	A A B	A B B	A B B
modifiers)	5-6	A B B	A B C	B C C	B C D	B B C	A B C	B C C
	7-8	B C E	A C D	B D E	D E F	C C E	B C D	C E E
	9	B D F	B D E	C E F	E E F	D E F	C D E	D F F

FIRING MODIFIERS

+2d10 Per unit/gun if aimed fire versus stationary target (firer must not have moved this turn or last, target not moved this turn
+1D10 per stand firing with rifles or carbines and stationary this turn
-1D10 per unit/gun if target in solid cover (eg walls, earthworks, buildings or dense woods)
-1D10 per unit/gun if target moved over half this turn (artillery turning doesn't count)
-1D10 per unit/gun if firer moved more than half (artillery turning does count)
-1D10 if firer is disordered
-1D10 per stand unable to fire (no line of sight) or lost (eg regiment down to 3 stands = -2 dice)
Regardless of modifiers, the firing unit may always roll at least 1 die.

Quick Reference Sheet - Firing (2)

Artillery	Short	Medium	Long
Light shot or shell	0 - 8	over 8 - 16	over 16 - 24
Light canister	0 - 4	-	-
Med shot or shell	0 - 12	over 12 - 20	over 20 - 32
Medium Canister	0 - 6	-	-
Hvy Shot or shell	0 - 16	over 16 - 28	over 28 - 40
Heavy Canister	0 - 8	-	-

Artillery Firing Table

Target formation and range (canister, short, medium, long)

Artillery unit's experience	best roll	Infantry Line C S ML	Inf column or square C S ML	Inf Extend. Line C S ML	Skirmisher Dis. Cav C S ML	Mounted Cavalry C S ML	Unlimb'd Artillery C S ML	Limbered Artillery C S ML
	0-2	B A B C	A A B B	C C D E	C B C D	B A B C	A A C D	A A B C
Green	3-4	C B C D	B A C C	D D E F	D C D E	C B D E	B B D E	B B C D
	5-6	D C D E	C B D E	E E E F	E D E F	D C E F	C C E F	C C D E
	7-8	E D E F	D C E F	E F F F	F E F F	E D F F	D D F F	D D E F
	9	F E F F	E D F F	F F F F	F F F F	F E F F	E E F F	E E F F
Trained	0-2	B A B C	A A A B	B A B C	B A B C	A A B C	A A B C	A A B C
	3-4	B B B D	B A A C	C B C D	C B C D	B A C D	A A C D	B B C D
	5-6	C B C E	B A B D	D C D E	D C D E	C B C D	B B D E	C C D E
	7-8	D C D F	C B C E	E D E F	E D E F	D C D E	C C E F	D D E F
	9	E D E F	D C D F	F E F F	F E F F	E D E F	D D F F	E E E F
Experi enced	0-2	A A B B	A A A A	A A A B	A A A B	A A A A	A A A A	A A B B
	3-4	B A B C	A A A B	B B B C	B B B C	B A B B	A A B C	B B C D
	5-6	C B D D	B A B C	C C D D	C C C D	B B C D	B B C D	C C D E
	7-8	D C E F	C B C D	D D E F	D D E F	C C D E	C C D E	D D E F
	9	E D F F	D C D E	E E F F	E E F F	D D E F	D D E F	E E F F
Veteran	0-2	A A A B	A A A A	A A A A	A A A A	A A A A	A A A A	A A A B
	3-4	B A B C	A A A B	A A B B	A A B B	A A A B	A A B B	A A B C
	5-6	B B C D	B A B B	B B C C	B B C C	B A B C	B B C E	B B C D
	7-8	C B D E	B A B C	C C D E	C C D E	C B C D	C C D E	C C D E
	9	D C E F	C B C D	D D E F	D D E F	D C D E	D D E F	D D E F
Elite	0-2	A A A B	A A A A	A A A A	A A A A	A A A A	A A A A	A A A A
	3-4	A A A C	A A A B	A A A B	A A A B	A A A A	A A A B	A A A B
	5-6	B A B D	A A B B	B B B C	B B B C	A A B B	A A B C	A A B C
	7-8	C B C E	B A B C	B C C D	B C C D	B B C C	B B C D	B B C D
	9	D B D F	C B B D	C D D E	C D D E	C C D E	C C D E	C C D E

Firing Results Key

Result	Musketry vs inf or cavvs	Artillery	Artillery vs Infantry or Cavalry Canister	Artillery vs Infantry or Cavalry Shot or Shell	Artillery vs Artillery
A	2 stands*	3 crew*	3 stands*	2 stands*	gun destroyed, 1crew, morale test*
B	2 stands*	2 crew*	2 stands*	1 stand, morale test*	2 crew*
C	1 stand	1 crew	1 stand Shaken	1 crew	
D	Disordered	Shaken	Shaken	Disordered	Shaken
E	Morale test*	Morale test*	Disordered	Morale test*	Morale test*
F	No effect	No effect	Morale test*	No effect	No effect

* Take morale tests immediately, always take morale test if 2 or more stands/crew killed in one phase

Quick Reference Sheet - Charging (1)

1. Check charge distance 2. Check attacker's resolve (take charge morale check if required) 3.Check defenders resolve (take charge morale test if required) 4. If appropriate, move charging unit into contact.

ATTACKER'S RESOLVE TABLE

Morale	Experience/Roll	0	1	2	3	4	5	6	7	8	9
IMPETUOUS	Green	A	A	A	A	A	K	K	M	N	N
	Trained	A	A	A	G	J	K	L	M	M	N
	Experienced	A	A	C	G	J	J	L	L	M	N
	Veteran	A	A	B	E	G	I	K	L	M	M
	Elite	A	A	B	B	E	G	H	K	L	M
CAUTIOUS	Green	A	A	G	K	L	L	M	M	N	N
	Trained	A	B	G	G	G	K	L	M	M	N
	Experienced	A	D	G	G	I	J	K	L	M	N
	Veteran	A	D	F	G	G	H	J	L	M	M
	Elite	A	B	D	E	E	G	H	K	L	M
STEADY	Green	A	A	A	G	K	K	M	M	N	N
	Trained	A	A	C	G	J	K	L	M	N	N
	Experienced	A	B	E	F	G	I	K	L	M	N
	Veteran	A	B	D	E	F	G	H	J	M	N
	Elite	A	B	B	B	B	D	G	H	I	M
CONFIDENT	Green	A	A	A	B	C	G	J	K	M	N
	Trained	A	A	B	C	F	G	H	L	M	N
	Experienced	A	B	B	E	F	F	G	K	M	N
	Veteran	A	B	B	D	D	E	F	H	I	L
	Elite	A	B	B	D	D	D	E	G	H	L
ARROGANT	Green	A	A	A	A	A	A	C	K	M	N
	Trained	A	A	A	A	C	C	K	M	M	N
	Experienced	A	A	A	B	B	E	H	L	M	M
	Veteran	A	A	A	B	B	B	E	H	L	M
	Elite	A	A	A	B	B	B	C	D	H	M

ATTACKER'S RESOLVE TABLE KEY

Result	Infantry	Cavalry
A	Charge home	Charge home
B	Attack at walk, no charge bonus	Charge home at trot., no charge bonus
C	Close to short range and fire. No charge	Charge disordered, no charge bonus, penalty for disordered
D	Close to medium range and fire. No charge	Charge disordered, no charge bonus, penalty for disordered
E	Close to long range and fire. No charge.	Move 1/2 distance, become disordered
F	Take charge morale test	Move 1/2 distance, become disordered
G	Take charge morale test	Halt, do not charge
H	Pull back 1 inch and fire (if still in range)	Halt, do not charge
I	Halt and fire at close range - rout if countercharged	Break off charge and retire 2 inches
J	Halt and fire at med. range - rout if countercharged	Break off charge and retire 4 inches
K	Halt and fire at long range - rout if countercharged	Take charge morale test
L	Retire 2 inches facing the enemy	Take charge morale test
M	Retire disordered 2 inches NOT facing the enemy	Rout
N	Rout	Rout

Quick Reference Sheet - Charging (2)

DEFENDER'S RESOLVE TABLE

	ORDERS	ASSAULT				ADVANCE				ENGAGE				HOLD				RETIRE			
Morale	**Experience**	**0-3**	**4-5**	**6-8**	**9**	**0-3**	**4-5**	**6-8**	**9**	**0-3**	**4-5**	**6-8**	**9**	**0-3**	**4-5**	**6-8**	**9**	**0-3**	**4-5**	**6-8**	**9**
IMPETUOUS	Green	Q	R	Y2	Z	Q	R	T	Z	Q	W	Y1	Z	R	Y1	Y2	Z	R	Y2	Y3	Z
	Trained	Q	R	Y1	Y2	R	T	Y1	Y3	R	V	Y1	Z	S2	Y1	Y2	Z	S3	Y2	Y3	Z
	Experienced	Q	T	X	Y1	R	S3	W	Y2	S1	S2	V	Y3	S2	Y1	Y1	Z	S3	Y1	Y2	Z
	Veteran	R	S2	T	Y1	S1	S1	S3	Y1	S2	S2	U	Y2	S3	T	W	Y3	U	U	V	Z
	Elite	R	R	S1	T	S1	S2	S3	U	S1	S1	S2	V	S2	S3	T	X	S2	U	U	Y1
CAUTIOUS	Green	Q	S2	T	W	S2	T	V	W	S2	V	W	X	S3	T	W	Y3	W	Y1	Y2	Z
	Trained	R	S1	T	W	S1	V	W	W	S2	V	V	X	S2	T	V	Y2	T	W	Y2	Y3
	Experienced	R	S1	U	V	S1	S3	T	V	S1	T	U	W	S2	S3	T	X	S2	T	X	Y1
	Veteran	S1	T	U	U	S1	S2	T	V	S1	S2	T	V	S1	S2	S3	W	S2	S3	T	X
	Elite	S1	U	U	V	S1	S2	S3	V	S1	S2	S3	V	S1	S1	S2	W	S1	S2	S3	W
STEADY	Green	R	T	T	Y1	R	S2	T	Y2	S2	T	T	Y2	T	T	T	Y3	T	W	Y2	Z
	Trained	R	T	W	Y1	S2	S2	S3	Y1	S3	T	U	Y1	S3	S3	T	Y2	V	W	Y1	Y2
	Experienced	S1	S3	T	U	S1	S2	S2	U	S1	S3	T	V	S2	S2	S3	W	S3	V	W	Y1
	Veteran	S1	S2	S3	U	R	S1	S2	U	S1	S2	S3	U	S1	S1	S2	V	S2	T	U	W
	Elite	U	U	U	V	R	R	S1	V	S1	S1	S2	V	S1	S1	S1	V	S2	S3	T	V
CONFIDENT	Green	Q	R	S3	T	Q	R	T	W	R	S3	T	W	R	T	W	Y1	S2	S3	W	Y2
	Trained	R	S2	T	U	R	R	S2	U	R	S2	S3	U	S2	S3	T	V	S2	S3	V	W
	Experienced	S1	S2	S3	T	S1	S2	S3	T	S2	S2	S3	U	S2	S2	S3	V	S2	S3	T	V
	Veteran	S1	S2	U	U	S1	S1	S2	U	S1	S1	S2	U	S2	S2	S2	U	S2	S2	S3	V
	Elite	S1	S1	S2	U	S1	S1	S1	U	S1	S1	S1	U	S1	S1	S2	U	S2	S2	S2	U
ARROGANT	Green	Q	Q	R	T	Q	R	Y1	Y2	R	R	S3	W	R	S3	T	W	T	T	W	Y1
	Trained	Q	R	R	S3	Q	T	T	T	S2	S3	T	U	S3	S3	S3	V	S3	S3	V	W
	Experienced	R	S2	T	U	R	R	T	U	S1	S2	S3	U	S2	S2	S3	V	S2	S3	T	V
	Veteran	R	S1	U	U	R	R	S1	U	S1	S1	S2	U	S1	S2	S3	U	S2	S2	S2	V
	Elite	S1	S1	U	U	R	R	R	U	S1	S1	S1	U	S1	S1	S1	U	S1	S2	S3	V

Result	Infantry	Dismounted Cavalry	Mounted Cavalry	Artillery
Q	Counter-charge	Fire full eff. retire 1/2 move	Counter-charge	Fire double canister
R	Fire at short range, then fight	Retire full move and remount	Counter-charge	Fire at short range effect
S1	Fire med range, then fight	Fire 1/2 eff., retire full move	Receive charge stationary	Fire at medium range effect
S2	Fire at long range, then fight	Retire full move, disordered	Receive charge stationary	Fire at long range effect
S3	Stand, fire if not charged	Morale Test	Receive charge stationary	Limber gun, retire 1/2 move
T	Morale Test	1 std flee, rest fight if charged	Counter-charge disordered	Morale Test
U	Pull back 1/2 move and fire	2 std flee, rest fight if charged	Morale Test	Defend gun, fight if charged
V	Pull back full move and fire	Ret 1 mve facing , horses bolt enemy	Retire 1/2 move, face	Abandon gun, crew flee 1/2
W	Retire not facing the enemy	As above, but NOT facing	Retire full move, face enemy	Spike gun and flee 1/2 move
X	Retire disordered, not facing	Shaken, cannot move or fight	Retire full move not facing	Surr. if chgd or spike and rout
Y1	1 std flees, rest stand & fight	Rout	As X, plus disordered	Spike gun and rout
Y2	2 stds flee, rest stand & fight	Rout	1 std flees, rest stand & fight	Rout, leaving gun
Y3	3 stds flee, rest stand & fight	Rout	2 stds flee, rest stand & fight	Rout, leaving gun
Z	Unit routs	Rout	Rout	Rout, leaving gun

Quick Reference Sheet - Charge Morale Tests

Use this table only for Morale Tests required by Attacker or Defender Resolve Table

CHARGE MORALE TABLE

MORALE	EXPERIENCE	ATTACKER				DEFENDER WITH ASSAULT ORDER				DEFENDER WITH ADVANCE ORDER				DEFENDER WITH ENGAGE ORDER				DEFENDER WITH HOLD ORDER				DEFENDER WITH RETIRE ORDER			
		0-3	4-5	6-8	9	0-3	4-5	6-8	9	0-3	4-5	6-8	9	0-3	4-5	6-8	9	0-3	4-5	6-8	9	0-3	4-5	6-8	9
IMPETUOUS	Green	A	A	G	J	B	C	E	F	B	C	E	F	B	C	E	J	C	E	J	L	E	F	I	L
	Trained	A	C	H	K	B	D	J	F	B	D	E	F	B	E	F	H	E	G	I	K	H	I	K	L
	Experienced	A	C	D	F	A	B	D	G	B	C	D	E	C	D	E	F	D	H	I	J	H	I	I	J
	Veteran	A	F	F	H	A	B	C	D	A	B	F	G	C	D	D	E	D	G	H	I	G	I	J	K
	Elite	A	B	C	I	A	C	D	G	A	B	D	F	A	B	F	G	D	F	G	H	G	H	I	K
CAUTIOUS	Green	A	B	E	H	B	C	G	K	C	E	J	K	E	G	H	I	E	I	J	K	H	I	K	L
	Trained	A	D	G	K	B	E	F	H	C	G	H	J	G	H	H	I	E	F	H	I	I	J	K	L
	Experienced	B	F	H	I	B	D	E	F	B	D	F	G	D	F	H	I	F	G	H	I	G	H	I	K
	Veteran	B	C	F	H	C	D	F	G	B	C	F	G	D	E	F	G	D	E	F	G	G	I	J	K
	Elite	B	F	F	I	A	C	D	F	A	D	E	G	B	C	D	F	C	D	E	F	F	G	I	K
STEADY	Green	A	E	G	L	B	C	D	G	B	E	F	G	B	D	I	K	E	I	J	K	H	I	K	L
	Trained	A	D	H	K	C	E	H	J	B	C	D	E	C	E	H	J	E	H	I	J	H	J	K	L
	Experienced	B	C	F	H	C	D	F	I	C	D	G	H	D	F	G	H	E	F	G	H	I	I	J	L
	Veteran	B	F	H	I	B	C	D	F	C	D	E	F	D	F	F	G	D	G	H	I	G	H	I	J
	Elite	B	F	H	I	A	D	F	H	C	D	F	G	D	E	F	F	D	E	F	G	G	G	H	I
CONFIDENT	Green	A	B	D	G	B	C	E	F	B	E	G	H	C	E	F	H	C	E	G	H	H	I	K	L
	Trained	A	C	G	H	C	D	F	G	B	D	G	I	C	D	F	G	D	G	H	I	E	I	J	K
	Experienced	B	C	F	I	A	C	D	F	B	C	D	F	B	C	D	F	D	E	G	H	E	H	I	J
	Veteran	B	C	F	F	A	B	D	F	A	C	D	E	B	C	D	E	D	E	G	I	E	G	I	J
	Elite	B	C	D	F	A	B	C	D	A	B	C	D	A	B	D	F	D	F	G	H	D	G	H	I
ARROGANT	Green	A	B	B	E	B	C	E	F	B	C	F	I	C	E	F	G	B	E	E	H	E	H	I	K
	Trained	A	B	E	F	B	D	F	G	B	D	F	H	C	D	E	F	D	E	G	H	H	I	J	K
	Experienced	A	A	C	F	A	B	D	G	A	C	D	E	B	C	D	E	C	D	E	F	G	H	I	J
	Veteran	A	B	F	F	A	A	C	D	A	B	D	E	A	D	E	F	C	D	F	G	D	G	H	I
	Elite	A	C	F	F	A	A	A	F	A	B	C	D	A	B	D	E	A	C	D	F	C	D	G	I

Charge Morale Test Results Key

Result	Attacker Infantry	Attacker cavalry	Defender Infantry	Defender cavalry	Def. dismtd cav.	Defender Artillery
A	Charge to contact	Charge to contact	Counter-charge	Counter-charge	Fire short range	Fire double canister
B	Attack disordered	Attack disordered	Stand, fire if not charged	Counter-charge disordered	Fire long range, mount ,retire 4in	Ret by recoil 1 inch
C	Halt, fire short rge	Attack disordered	Fire at close range, fight if charged	Receive at halt	Mount up and retire full move	Crew retreat 4in, abandon gun
D	Attack falters - test resolve again	Refuse to charge, halt	Fire at half effect using 2D10, fight	Receive at halt	Mount up and retire half move	Crew pinned, surrender if attacked
E	Fall back 4 inches	Fall back 6 inches	Hold orders changed to retire	About turn and retire for 8 inches	Retire half move on foot	Crew pinned - will be killed if attacked
F	Fall back 6 inches	Fall back 8 inches	Fall back 4 inches	Turn, retire 8in disordered	Retire full move on foot	Crew abandons gun and routs
G	One stand flees, remainder fire	Assault orders changed to advance	Fall back 6 inches	One stand flees-remainder fights	Horse's bolt, retire half a move on foot	Crew abandons gun and routs
H	Two stands flee, remainder fire	Turn about and retire a full move	One stand flees, remainder fights	Two stands flee, remainder fights	Horse's bolt, retire a full move on foot	Crew abandons gun and routs
I	Halt and fire at half effect using 2D10	Feint.. Halt halfway to enemy	Two stands flee, remainder fights	Retire a full move	Panic and rout	Crew abandons gun and routs
J	Assault orders changed to Advance	Feint. Halt halfway to enemy, disordered	Three stands flee, remainder fights	Evade full move and scatter - 2 turns to reform	Panic and rout	Crew abandons gun and routs
K	Panic and rout	Panic and halt shaken	Panic and rout	Panic and rout	Panic and rout	Crew abandons gun and routs
L	Panic and rout	Panic and Rout	Panic and rout	Panic and rout	Panic and rout	Crew abandons gun and routs

Quick Reference Sheets - Close Combat

1. Determine no of attacker's dice and roll. Check modifiers to see how many to roll and apply.
2. Select best result(s),apply to Attacker's Melee Resolution Table and Melee Results Key
3. Before removing casualties, Defender uses Defender's Melee Resolution Table and Melee Results Key
4. Remove all casualties, move units pushed back etc. 5. Conduct melee morale test

Usually single best result is applied to Resolution Table, but see modifiers below

Melee Modifiers

Charging , count results of 2 dice
Attacking flank: count results of 3 dice
Cav vs Inf square of 6+ stands, -1 die
Disordered (att or def) roll -1 die

Counter-charging cancels charger's bonus
Behind defences, defender rejects 2 of attacker's rolls
Inf square v Cav may cancel enemy charge bonus OR roll 1 extra die on Defender Resolution

For each stand below full strength - opponent can discard that number of rolls
Extended line versus line or column, roll 1 fewer dice

Attacker (Defender) Melee Resolution	**Die roll**	**Inf v Inf**	**Inf v Cav**	**Cav v Inf**	**Cav v Cav**	**Inf v Art**	**Cav v Art**
	0-2	B (B)	D (D)	B (B)	B (A)	A	A
Green	3-4	D (C)	E (E)	C (C)	D (B)	B	A
(1D10)	5-6	E (D)	G (G)	D (E)	E (D)	C	B
	7-8	F (E)	H (H)	G (G)	F (E)	E	C
	9	H (F)	I (I)	J (I)	J (F)	E	D
Trained	0-2	A (A)	B B()	B A()	B (A)	A	A
(2D10)	3-4	C (C)	D D()	C (B)	C (B)	B	A
	5-6	D (E)	E (E)	D (D)	E (C)	C	B
	7-8	E (F)	G (G)	D (E)	F (E)	D	B
	9	F (F)	H (H)	E (G)	G (F)	E	C
Experienced	0-2	A (A)	B (B)	A (A)	B (A)	A	A
(3D10)	3-4	B (B)	C (C)	B (B)	C (B)	A	A
	5-6	C (E)	D (E)	C (D)	D (C)	B	B
	7-8	D (F)	E (G)	D (E)	E (D)	C	C
	9	E (F)	F (H)	E (G)	F (B)	D	D
Veteran	0-2	A (A)	B (B)	A (A)	A (A)	A	A
(4D10)	3-4	A (B)	C (C)	A (B)	B (B)	A	A
	5-6	B (C)	D (D)	B (D)	C (B)	B	A
	7-8	D (D)	D (E)	C (E)	D (C)	B	D
	9	E (E)	E (G)	D (H)	E (D)	C	C
Elite	0-2	A (A)	B (B)	A (A)	A (A)	A	A
(4D10)	3-4	A (B)	C (C)	A (B)	B (B)	A	A
	5-6	B (C)	D (E)	B (C)	C (C)	A	A
	7-8	B (E)	D (G)	B (E)	C (E)	A	A
	9	C (G)	E (H)	C (H)	E (F)	C	C

A - 2 enemy stands killed. **B -** 1 enemy stand killed. **C -** 1 enemy stand flees. **D -** Enemy pushed back 4". **E -** Enemy test morale. **F -** Stalemate. **G -** Own unit pushed back 4". **H -** 1 of own stand killed. **I -** 2 of own stands killed. **J -** Own unit test morale.

Quick Reference Sheet - Morale Tests

Check morale when: Unit from own brigade routed or seen wiped out within 12"/To recover from shaken, disordered or routing/unit's brigade or higher officer killed within 12"/Unit has taken casualties this turn and not taken a morale test for any other reason.

Morale tests resulting from charging procedure (testing resolve) use the Charge Morale Table (p.144). All other morale tests are taken as follows. Roll 2D10 ('0' = 10) and apply modifiers. Final score lower than unit's Basic Factor (BF) = test passed, otherwise failed by number BF is exceeded by.

Add 1 to dice roll

Each stand lost by unit to date

If the Brigade Commander has been lost this turn

Raw Trained or Raw Green troops under fire for 1st time

Currently disordered or shaken

If there are no friendly units within 4 inches

Any enemy unit within 4 inches

Fired on or attacked from flank or rear

EACH friendly unit routing, retiring or destroyed in 5 inches

Unit pushed back in melee

Subtract 1 from dice roll

If unit is at full strength

Brigade commander within 5 inches
If the Divisional Cdr is within 5 inches

If the Army or Corps Cdr is within 6 inches
If there are friendly troops within 4 inches

Unit has inflicted casualties on enemy this turn

If unit is in cover
Any routing or retiring enemy within 4 inches

If unit has been victorious in melee

Morale Table

		Assault Order failed by					Advance Order failed by					Engage Order failed by					Hold Order failed by					Retire Order failed by				
Morale	Experience	1-4	5-8	9-12	13-16	11-20	1-4	5-8	9-12	13-16	11-20	1-4	5-8	9-12	13-16	11-20	1-4	5-8	9-12	13-16	11-20	1-4	5-8	9-12	13-16	11-20
Impetuous BF9	Green	A	C	F	G	H	A	A	B	D	F	A	A	B	F	G	A	B	B	E	G	A	B	F	G	H
	Trained	A	B	C	F	H	A	B	D	F	G	A	B	F	G	H	B	E	F	G	H	B	F	G	H	I
	Experien'd	B	B	G	H	I	B	D	F	G	H	B	F	G	H	I	E	F	G	H	I	F	G	H	I	J
	Veteran	B	C	H	H	I	D	F	G	H	I	B	F	F	G	H	F	G	G	H	I	G	G	H	I	J
	Elite	B	C	C	H	I	F	G	H	I	J	B	F	G	G	H	G	H	H	I	J	H	H	I	J	J
Cautious BF8	Green	B	C	F	H	I	B	B	D	F	G	A	B	F	G	H	B	E	G	H	I	F	G	H	I	J
	Trained	B	F	G	H	I	B	D	F	G	H	B	F	G	H	I	E	G	H	H	I	G	H	I	I	J
	Experien'd	A	B	C	H	I	D	F	G	H	I	F	G	H	I	I	F	G	H	I	J	H	H	I	J	J
	Veteran	A	C	G	H	I	D	G	H	I	J	F	H	I	I	J	G	H	I	I	J	H	I	I	J	J
	Elite	B	C	G	H	I	G	H	I	I	J	H	H	I	J	J	E	G	H	I	J	I	I	J	J	J
Steady BF10	Green	A	A	B	F	G	A	B	D	G	H	B	F	G	H	I	A	E	F	G	H	B	G	H	I	J
	Trained	A	B	C	G	H	B	D	F	G	H	F	G	H	I	J	B	E	E	F	G	F	G	H	I	J
	Experien'd	B	C	F	H	I	B	F	G	H	I	F	F	G	H	I	E	F	G	H	I	G	H	H	I	J
	Veteran	B	C	G	H	I	A	B	F	G	H	A	B	F	G	H	E	G	H	I	J	H	H	I	I	J
	Elite	A	B	F	G	H	B	F	G	H	H	B	F	G	H	I	E	G	H	I	I	H	I	I	J	J
Confident BF11	Green	A	A	B	F	G	A	B	D	F	G	A	A	B	B	F	A	B	B	E	F	B	B	F	G	H
	Trained	A	A	C	G	H	B	D	F	G	H	A	B	F	F	G	A	B	E	F	G	B	F	G	H	I
	Experien'd	A	B	B	C	F	B	F	G	H	I	A	B	F	G	H	B	E	F	G	H	F	G	H	I	J
	Veteran	A	B	C	F	G	B	D	F	G	H	B	B	F	G	H	E	F	G	H	I	G	G	H	I	J
	Elite	A	C	F	G	H	D	F	G	H	H	B	F	G	H	I	F	G	H	I	J	G	H	I	I	J
Arrogant BF12	Green	A	A	A	B	C	A	A	A	B	D	A	A	B	B	F	A	B	E	F	G	A	B	F	G	H
	Trained	A	B	B	C	F	A	A	B	D	F	A	B	B	F	G	A	E	F	G	H	A	F	G	H	I
	Experien'd	A	B	C	F	G	A	B	D	F	G	A	B	F	G	H	A	B	E	F	G	A	G	H	I	J
	Veteran	A	C	F	G	H	B	D	F	G	H	A	F	G	H	I	A	B	B	E	F	A	H	I	I	J
	Elite	B	C	F	G	H	D	F	G	H	I	A	G	H	I	J	A	B	E	F	G	A	I	I	J	J

Infantry or Cavalry: A - Disordered, must halt. **B** - Shaken. **C** - Assault order changes to Engage. **D** - Advance orders change to Hold. **E** - Hold orders to change. **F** - Retire 2", not facing enemy. **G** - Retire 4", not facing enemy. **H** - Retire 6", not facing enemy. **I** - Rout as mob. **J** - Rout scattered and disordered.

Deployed Artillery: A - Shaken. **B** - H orders change to Retir. **C** - Limber and retire 2". **D** - Limber and retire 4". **E** - Abandon gun, run to nearest friends. **F** - Spike gun, retire 6". **G** - Spike gun, surrender if enemy in 4". **H** - Gun captured, crew surr. if enemy in 4". **I** - Crew rout, gun captured. **J** - /.

Limbered Artillery: A - Shaken. **B** - H orders change to Retire. **C** - Retire 4". **D** - Retire 8". **E** - Ride towards nearest friends. **F** - Retire 4", gun overturned. **G** - Captured if enemy in 4" or rout. **H** - Rout. **I** - Rout. **J** - Rout.

Routers: A - Rally but shaken to reform. **B** - Rally but disordered. **C** - Rout continues. **D** - Rout continues. **E** - Rout continues. **F** - Rout continues. **G** - Rout continues. **H** - Rout continues. **I** - Dispersed, remove. **J** - Dispersed, remove.

Order Markers

ASSAULT	ADVANCE	ENGAGE	HOLD	RETIRE	ROUT	SHAKEN	DOUBLE
ASSAULT	ADVANCE	ENGAGE	HOLD	RETIRE	ROUT	SHAKEN	DOUBLE
ASSAULT	ADVANCE	ENGAGE	HOLD	RETIRE	ROUT	SHAKEN	DOUBLE
ASSAULT	ADVANCE	ENGAGE	HOLD	RETIRE	ROUT	SHAKEN	DOUBLE
ASSAULT	ADVANCE	ENGAGE	HOLD	RETIRE	ROUT	SHAKEN	DOUBLE
ASSAULT	ADVANCE	ENGAGE	HOLD	RETIRE	ROUT	SHAKEN	DOUBLE
ASSAULT	ADVANCE	ENGAGE	HOLD	RETIRE	ROUT	SHAKEN	DOUBLE
ASSAULT	ADVANCE	ENGAGE	HOLD	RETIRE	ROUT	SHAKEN	DOUBLE
ASSAULT	ADVANCE	ENGAGE	HOLD	RETIRE	ROUT	SHAKEN	DOUBLE
ASSAULT	ADVANCE	ENGAGE	HOLD	RETIRE	ROUT	SHAKEN	DOUBLE
ASSAULT	ADVANCE	ENGAGE	HOLD	RETIRE	ROUT	SHAKEN	DOUBLE
ASSAULT	ADVANCE	ENGAGE	HOLD	RETIRE	ROUT	SHAKEN	DOUBLE
ASSAULT	ADVANCE	ENGAGE	HOLD	RETIRE	ROUT	SHAKEN	DOUBLE
ASSAULT	ADVANCE	ENGAGE	HOLD	RETIRE	ROUT	SHAKEN	DOUBLE
ASSAULT	ADVANCE	ENGAGE	HOLD	RETIRE	ROUT	SHAKEN	DOUBLE
ASSAULT	ADVANCE	ENGAGE	HOLD	RETIRE	ROUT	SHAKEN	DOUBLE
ASSAULT	ADVANCE	ENGAGE	HOLD	RETIRE	ROUT	SHAKEN	DOUBLE
ASSAULT	ADVANCE	ENGAGE	HOLD	RETIRE	ROUT	SHAKEN	DOUBLE
ASSAULT	ADVANCE	ENGAGE	HOLD	RETIRE	ROUT	SHAKEN	DOUBLE
ASSAULT	ADVANCE	ENGAGE	HOLD	RETIRE	ROUT	SHAKEN	DOUBLE
ASSAULT	ADVANCE	ENGAGE	HOLD	RETIRE	ROUT	SHAKEN	DOUBLE
ASSAULT	ADVANCE	ENGAGE	HOLD	RETIRE	ROUT	SHAKEN	DOUBLE
ASSAULT	ADVANCE	ENGAGE	HOLD	RETIRE	ROUT	SHAKEN	DOUBLE
ASSAULT	ADVANCE	ENGAGE	HOLD	RETIRE	ROUT	SHAKEN	DOUBLE
ASSAULT	ADVANCE	ENGAGE	HOLD	RETIRE	ROUT	SHAKEN	DOUBLE
ASSAULT	ADVANCE	ENGAGE	HOLD	RETIRE	ROUT	SHAKEN	DOUBLE
ASSAULT	ADVANCE	ENGAGE	HOLD	RETIRE	ROUT	SHAKEN	DOUBLE
ASSAULT	ADVANCE	ENGAGE	HOLD	RETIRE	ROUT	SHAKEN	DOUBLE
ASSAULT	ADVANCE	ENGAGE	HOLD	RETIRE	ROUT	SHAKEN	DOUBLE
ASSAULT	ADVANCE	ENGAGE	HOLD	RETIRE	ROUT	SHAKEN	DOUBLE
ASSAULT	ADVANCE	ENGAGE	HOLD	RETIRE	ROUT	SHAKEN	DOUBLE
ASSAULT	ADVANCE	ENGAGE	HOLD	RETIRE	ROUT	SHAKEN	DOUBLE
ASSAULT	ADVANCE	ENGAGE	HOLD	RETIRE	ROUT	SHAKEN	DOUBLE
ASSAULT	ADVANCE	ENGAGE	HOLD	RETIRE	ROUT	SHAKEN	DOUBLE
ASSAULT	ADVANCE	ENGAGE	HOLD	RETIRE	ROUT	SHAKEN	DOUBLE
ASSAULT	ADVANCE	ENGAGE	HOLD	RETIRE	ROUT	SHAKEN	DOUBLE
ASSAULT	ADVANCE	ENGAGE	HOLD	RETIRE	ROUT	SHAKEN	DOUBLE
ASSAULT	ADVANCE	ENGAGE	HOLD	RETIRE	ROUT	SHAKEN	DOUBLE
ASSAULT	ADVANCE	ENGAGE	HOLD	RETIRE	ROUT	SHAKEN	DOUBLE
ASSAULT	ADVANCE	ENGAGE	HOLD	RETIRE	ROUT	SHAKEN	DOUBLE
ASSAULT	ADVANCE	ENGAGE	HOLD	RETIRE	ROUT	SHAKEN	DOUBLE
ASSAULT	ADVANCE	ENGAGE	HOLD	RETIRE	ROUT	SHAKEN	DOUBLE
ASSAULT	ADVANCE	ENGAGE	HOLD	RETIRE	ROUT	SHAKEN	DOUBLE
ASSAULT	ADVANCE	ENGAGE	HOLD	RETIRE	ROUT	SHAKEN	DOUBLE
ASSAULT	ADVANCE	ENGAGE	HOLD	RETIRE	ROUT	SHAKEN	DOUBLE
ASSAULT	ADVANCE	ENGAGE	HOLD	RETIRE	ROUT	SHAKEN	DOUBLE
ASSAULT	ADVANCE	ENGAGE	HOLD	RETIRE	ROUT	SHAKEN	DOUBLE
ASSAULT	ADVANCE	ENGAGE	HOLD	RETIRE	ROUT	SHAKEN	DOUBLE
ASSAULT	ADVANCE	ENGAGE	HOLD	RETIRE	ROUT	SHAKEN	DOUBLE
ASSAULT	ADVANCE	ENGAGE	HOLD	RETIRE	ROUT	SHAKEN	DOUBLE
ASSAULT	ADVANCE	ENGAGE	HOLD	RETIRE	ROUT	SHAKEN	DOUBLE
ASSAULT	ADVANCE	ENGAGE	HOLD	RETIRE	ROUT	SHAKEN	DOUBLE
ASSAULT	ADVANCE	ENGAGE	HOLD	RETIRE	ROUT	SHAKEN	DOUBLE